NASA SP-340

THE ATMOSPHERE OF TITAN

The proceedings of a workshop held at
Ames Research Center, July 25 through 27, 1973

Edited by

Donald M. Hunten
Kitt Peak National Observatory

Prepared by NASA Ames Research Center

Scientific and Technical Information Office   1974
NATIONAL AERONAUTICS AND SPACE ADMINISTRATION
Washington, D.C.

TITAN ATMOSPHERE WORKSHOP

Donald M. Hunten, Kitt Peak National Observatory, Chairman

Jacques E. Blamont
    Université de Paris

A. Lyle Broadfoot
    Kitt Peak National Observatory

John J. Caldwell
    Princeton University

Robert E. Danielson
    Princeton University

John S. Lewis
    Mass. Institute of Technology

Thomas R. McDonough
    Cornell University

David Morrison
    University of Hawaii

James B. Pollack
    Ames Research Center

Carl Sagan
    Cornell University

Darrell F. Strobel
    Naval Research Laboratory

Nicole Tabarié
    Service d' Aéronomie du CNRS

Laurence M. Trafton
    University of Texas

Joseph Veverka
    Cornell University

OTHER CONTRIBUTORS

F. H. Briggs
    Cornell University

T. A. Croft
    Stanford University

F. C. Gillett
    Kitt Peak National Observatory

D. H. Herman
    NASA Headquarters

S. I. Rasool
    NASA Headquarters

R. L. Younkin
    Jet Propulsion Laboratory

AMES RESEARCH CENTER

L. R. Boyce

L. Colin

W. L. Jackson

B. D. Padrick

PREFACE

The Uniqueness of Titan

Titan offers a unique opportunity in solar system exploration. It is the smallest known body with an atmosphere. In terms of spacecraft entry dynamics, it has the most accessible atmosphere in the solar system. It has dark reddish clouds which are probably composed of organic compounds, falling continually to the surface. It has the highest ratio of methane to hydrogen of all known reducing atmospheres, making an environment in some respects like that of the primitive Earth at the time of the origin of life. It probably has the only surface of all the bodies beyond Mars with atmospheres that entry spacecraft can reach. In terms of planetary rotation rate, Titan's atmospheric circulation may occupy a unique niche between the dynamics of Venus and the Earth. The surface temperature may be 150-200°K or warmer, and one model suggests an ocean of liquid methane and ammonia. While at the present this is the merest speculation, the presence of life on Titan is by no means out of the question. Hydrogen in the Titanian atmosphere must be rushing away to space at a high rate, perhaps producing an enormous gas ring or comet-like tail in the vicinity of Saturn. The escaping gas must be produced from the interior of Titan, perhaps arriving into the atmosphere through methane, ammonia, and water volcanoes.

A substantial improvement of our knowledge of Titan can be achieved with present instrumentation in the vicinity of the Earth -- by ultraviolet and infrared spectroscopy, infrared and microwave radiometry, and stellar occultations. Observational and modeling techniques that have been used to study the planets have just begun to be applied to Titan. Many important properties are accessible, and a considerable improvement in our knowledge of Titan can be expected in the near future. But a thorough characterization of the environment of Titan -- and, in particular, studies of the tantalizing questions of organic chemistry and surface morphology -- must await spacecraft investigations at or near Titan.

The Titan Atmosphere Workshop

On July 25 to 27, 1973, a workshop was convened at Ames Research Center under the chairmanship of D. M. Hunten. At the request of NASA Headquarters, the purpose was to define, as far as now possible, the atmosphere of Titan for use in the planning of future missions to that body. Titan's prominence is so recent that all the active workers could easily meet in a small room. More than half these people were actually present, and a good coverage of the appropriate disciplines was obtained.

It quickly became clear that less than half the required material was published, so rapid is the growth of Titanian studies. Nearly 2 of the 3 days were therefore devoted to review papers, presentation of new results, and discussion. The edited transcript of this symposium forms Chapter 2, the bulk of this document. Some additional material appears in Chapter 3. The Workshop's best attempt at an engineering model forms Chapter 1, along with a summary by the editor of Chapters 2 and 3. Our current knowledge is clearly inadequate for engineering purposes, but it was equally clear that a vast improvement is feasible with today's observational techniques. Half a dozen recommendations for immediate work, both at the telescope and in the laboratory, are therefore made in Chapter 4. Other recommendations, in Chapters 4 and 5, look further to the future; they include the impact of (and on) the Mariner Jupiter/Saturn (MJS) flyby missions, and possible probes to enter Titan's atmosphere.

The Workshop participants would like to record their thanks to the staff of Ames Research Center for the fine technical and editorial support of their work. John Niehoff of Science Applications, Inc. was responsible for the editorial production of the report. We thank the publishers of The Astronomical Journal, The Astrophysical Journal, Comments on Astrophysical Space Physics, Icarus, and Planets, Stars, and Nebulae Studied with Photopolarimetry for their permission to reprint copyrighted material.

CONTENTS

CONTENTS (CONTD)

LIST OF TABLES

Chapter 1

INTRODUCTION AND SUMMARY

1.1   RANGE OF ENGINEERING VARIABLES

Titan is the largest satellite of Saturn, sharing Saturn's heliocentric distance and year, but having a radius comparable to that of Mercury. Its rotation is usually assumed to be synchronous with its revolution about Saturn, with its pole perpendicular to its orbit (in Saturn's equatorial plane); thus its "day" is 16 days, its "seasons" result from an inclination of 27° (comparable to the Earth's), and its "year", 30 years long, has 675 "days". Its probable structure, comparable to that of the other large satellites of the outer planets (J I-IV, and Triton), is described by Lewis (1971), and includes a rocky, muddy core, a liquid (H_2O solution) mantle (most of the volume), possibly an ice crust, and an atmosphere. The atmosphere is considerably more massive than that of Mars, and far outweighs the wisps of gas possessed (at most) by any other satellite.

Many of the data are given with considerable reluctance, as a reading of Sections 1.1 and 1.2 will suggest. Nevertheless, they represent our state of knowledge in mid-1973 and the consensus of the Workshop. Table 1-1 shows the basic parameters, after Morrison's discussion. The mass is well determined, but the uncertainty in the radius reflects into the mean density, surface gravity, and measured albedo. It is not known whether the radius is that of the solid body or of a layer high in a dense atmosphere. The temperatures given represent equilibrium between thermal radiation and absorbed solar energy. The temperature T_{max} is probably not realized under an atmosphere as massive as Titan's; it is appropriate for bodies like Mars or the Moon.

Table 1-2 shows the wide range of possible compositions compatible with spectroscopic data. The baseline atmosphere outlined in the box represents Trafton's actual measurements and analysis. The molecular column densities also represent the volumetric proportions. If the atmosphere contains optically dense clouds, the total column densities could be much greater. The absolute minimum atmosphere is obtained by leaving out the H_2 and taking the lower error limit for CH_4: an abundance of 1.5 km-A and a surface pressure of 10 mb.

Table 1-2 illustrates the fact that the derived abundance of CH_4 depends on assumptions about other constituents, through their pressure broadening of spectral lines. Several sample atmospheres are shown, involving CH_4 with either H_2 or N_2; mixtures of all three could also be considered. The larger H_2 abundances, shown in parentheses, are not compatible with observation. N_2 is not observable, but could be present as a dissociation product of NH_3. A few cm-A of NH_3 could be present near the surface if it is as warm as 140°K. Other trace gases that could be considered are Ar^{40} from radioactive decay, and primordial rare gases such as neon. There is no more reason to suppose that Titan retained the latter than did Earth. Helium escapes so readily that an appreciable abundance is most unlikely.

1

Table 1-1. Titan Parameters

Radius	R	$= 2500 \pm 250$ km
Mass	M	$= 1.37 \pm .02 \times 10^{26}$ g
Distance from the Sun	a	$= 9.546$ AU
Orbital period	P	$= 15.95$ days
Mean density	ρ	$= 2.1 \pm 0.6$ g cm^{-3}
Acceleration of gravity	g	$= 146 \pm 30$ cm s^{-2}
Geometric visual albedo	P_v	$= 0.20 \pm .04$
Bolometric albedo	A_{bol}	$= 0.26 \pm .08$
Maximum sub-solar temperature	T_{max}	$= 116 \pm 3°$K
Effective temperature	T_e	$= 82 \pm 2°$K

Table 1-2. Possible Atmospheric Abundances (km-A)*

CH_4	H_2		N_2	
2.0 ± 0.5	0		0	
1.4	5.0		0	Baseline
1.0	15.0	or	1.5	
0.1	(220)	or	20.0	
0.01	(2500)	or	210.0	

* 1 cm-amagat (cm-A), also called cm-atm, is a column density of
2.687 x 10^{19} (Loschmidt's number) molecules per cm . Surface
partial pressure for 1 km-A is 10.4, 1.3, 18 mb (CH_4, H_2, N_2).

Pressures and temperatures are summarized in Table 1-3. The stratosphere is warmer than the effective temperature because the dark aerosol absorbs solar ultraviolet radiation (as does ozone in the Earth's stratosphere). Thermal emission from this warm region has been observed. Table 3 gives several candidate surfaces, lettered in such a way as to suggest that there are many intermediate possibilities. Surface (a) corresponds to a methane atmosphere, with or without some H_2. Surface (m) would require around 30 km-atm of N_2. The last one supposes a methane "ocean" containing about 5% of the mass of Titan. This medium could be liquid or gaseous, depending on the relation of the CH_4 critical point to the actual pressure-temperature situation.

Diurnal and seasonal variations of temperature are probably small; they could perhaps reach a range of 10°K for the thinnest atmospheres considered. Not only does the atmospheric gas carry heat around, there is a great deal of latent heat available from condensation and evaporation of CH_4.

Since the Workshop, several detailed models of the atmosphere have been prepared by Divine (1973).

Table 1-3.   Pressures and Temperatures

REGION	p (mb)	T (°K)
Entry	.001 - 1	150 ± 50
Candidate Surfaces		
(a)	20	80
(m)	~ 500	150 - 200
(z)	10^6	150 - 1000
A fine, red-brown aerosol is probably present up to at least the 1-mb level. It is irrelevant for engineering purposes. CH_4 cloud may be present at or below the 20-mb level, and haze at greater heights.		

1.2 SCIENTIFIC SUMMARY

D. M. Hunten

This section is a brief overview of Chapter 2, which contains the presentations made at the Workshop, with the discussion and some supplementary material. A few of the principal literature references are included, but references to Chapter 2 are mostly implied.

Composition of the Atmosphere

Methane absorptions are prominent in the spectrum, but their interpretation is complicated by the possibility of pressure broadening by other gases; moreover, the true surface may be hidden by clouds. If such gases are not important, the methane abundance is 2 km-A (Trafton, 1972b). With 20 km-A of N_2 (as an example), the methane comes to 0.1 km-A, and the surface pressure is 350 mb. The methane atmosphere thus gives the smallest mass, and a surface pressure of 20 mb. A "possible detection" of H_2, in the amount of 5 km-A, was reported by Trafton (1972a), and subsequent work seems to confirm this difficult result. Although H_2 is common in the outer solar system, its prominence on a body as small as Titan is surprising: the loss by escape is large, and a correspondingly large source must be found, as discussed below.

For other possible gases we must turn to models of the formation and interior composition of Titan, since observational data are lacking. The work of Lewis (1971) makes use of the mean densities of the satellites, along with the hypothesis that such bodies are accreted from the condensed fraction of the solar nebula. About 60% of the mass should be a solution of NH_3 in H_2O, and a further 5% should be CH_4. The presence of the latter in the atmosphere fits this picture, which however suggests that there is far more methane remaining in the interior or on the surface. No H_2O, and very little NH_3, should be in the atmosphere at the prevailing temperatures. Photolysis of NH_3 and CH_4, and escape of H_2, could produce N_2, as well as a considerable range of other compounds, most of which should condense into aerosols or on the surface. Thus, the best candidate for a third atmospheric gas seems to be N_2. Noble gases might have been retained in small quantities. Although their rarity on Earth does not encourage this idea, Cess and Owen (1973) have developed a greenhouse model based on a mixture of H_2 and Ne.

Trafton (1973) has reported the presence of additional absorptions that do not seem to be due to CH_4, though many of them are also present in the spectrum of Uranus. One may speculate that some of the photolysis products are responsible, but such possibilities are limited because most compounds must condense at the low temperatures of Titan's atmosphere. The best candidates are therefore C_2H_6, C_2H_4, and C_2H_2, and perhaps methylamine CH_3NH_2 if ammonia photolysis occurs. Not enough is known about the spectra of any of these compounds for an identification or rejection; even the possibility of weak CH_4 bands remains.

In the absence of an atmosphere, Lewis (1971) would predict a surface of water ice containing CH_4 as a clathrate and NH_3 in solution. At a depth of a few tens of kilometers this medium should be melted. In the extreme case of a very deep atmosphere, it is conceivable that melting could extend all the way to the surface; the liquid CH_4 would then float on the H_2O-NH_3 solution. If the (p, T) relation passed through the critical point of methane, the atmosphere would merge into the ocean with no phase change, and could be regarded as having a surface pressure of some 1000 bars (cf. Lewis and Prinn, 1973).

A more likely situation is a cold surface covered with a layer of photolysis products and their polymers. Such mixtures are usually dark in color, as in Titan.

Cloud and Haze

Two kinds of aerosol are to be expected in Titan's atmosphere: clouds of solid CH_4, and a photochemical haze (or smog). Veverka (1973) and Zellner (1973) have published observations of Titan's polarization, which can be obtained only to a phase angle of 6°. Despite this limitation, it is clear that negative polarization is absent, in striking contrast to observations of the Moon, Mars, Mercury, and many terrestrial solid surfaces. Positive polarization is shown by glassy surfaces and by atmospheric scattering. However, pure Rayleigh scattering by a gas is ruled out by the low albedo and the absence of a wavelength dependence. An absorbing aerosol is therefore suggested, and Veverka further suggests that it should be optically thick to hide the negative polarization from the surface. However, it is not at all obvious that ordinary planetary surfaces are a good model for Titan. Indeed, this body could well be covered by a glassy or tarry layer of photolysis products, which would give a positive polarization. If such a possibility is accepted, the atmosphere could still be optically thin (see postscript, p. 57).

Another line of evidence is the low ultraviolet albedo of Titan, observed at 2600 Å by Caldwell (1973) and above 3000 Å by Barker and Trafton (1973). The model by Danielson et al. (1973) shows that an absorbing aerosol is required at stratospheric heights; otherwise the atmosphere would be too bright. One possibility is CH_4 ice, darkened by radiation; but a photochemical smog seems far more likely. Indeed, the stratosphere must be heated by the ultraviolet energy absorbed, and is probably too warm for methane condensation. The evidence for photochemical haze is almost beyond question; CH_4 ice may also be present at low altitudes, but does not seem to be required.

Thermal Structure

Much of the current wave of interest in Titan is due to the striking results obtained by several observers in the thermal infrared, summarized by Morrison et al. (1972). The recent data of Gillett et al. (1973) show that still more is to be expected when full spectral coverage of the 8-30 micron region has been attained. For some time there was a remarkable unanimity in seeking the explanation through a greenhouse effect. This complacency was rudely shattered by Danielson and Caldwell, who pointed out that the available data were equally well satisfied by a model based on radiation from a warm stratosphere. Indeed, the latest infrared data, those of Gillett et al., are strongly in favor of a warm stratosphere, for they show a high brightness temperature at 8 μm, in the wing of the 7.7-micron band of CH_4. However, a significant greenhouse effect could still exist. Figure 1-1 shows two cartoons that compare and contrast the two types of models. The effective temperature

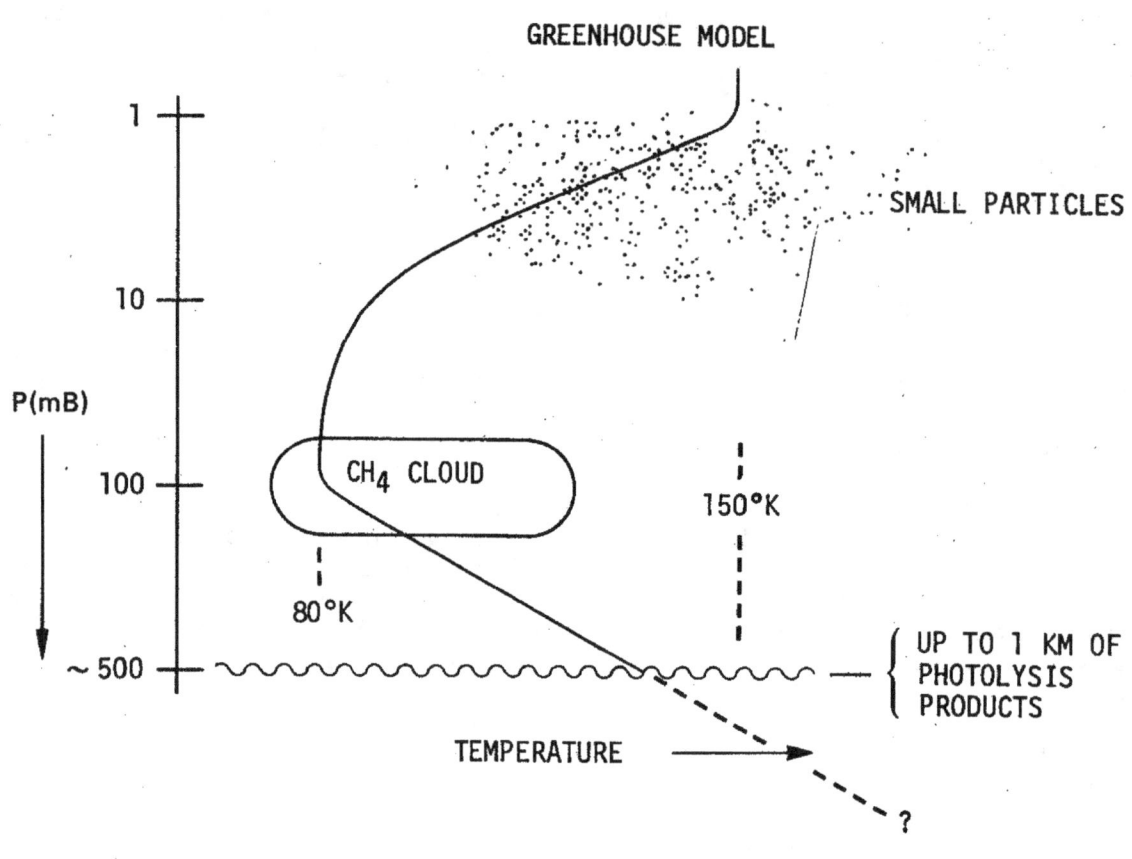

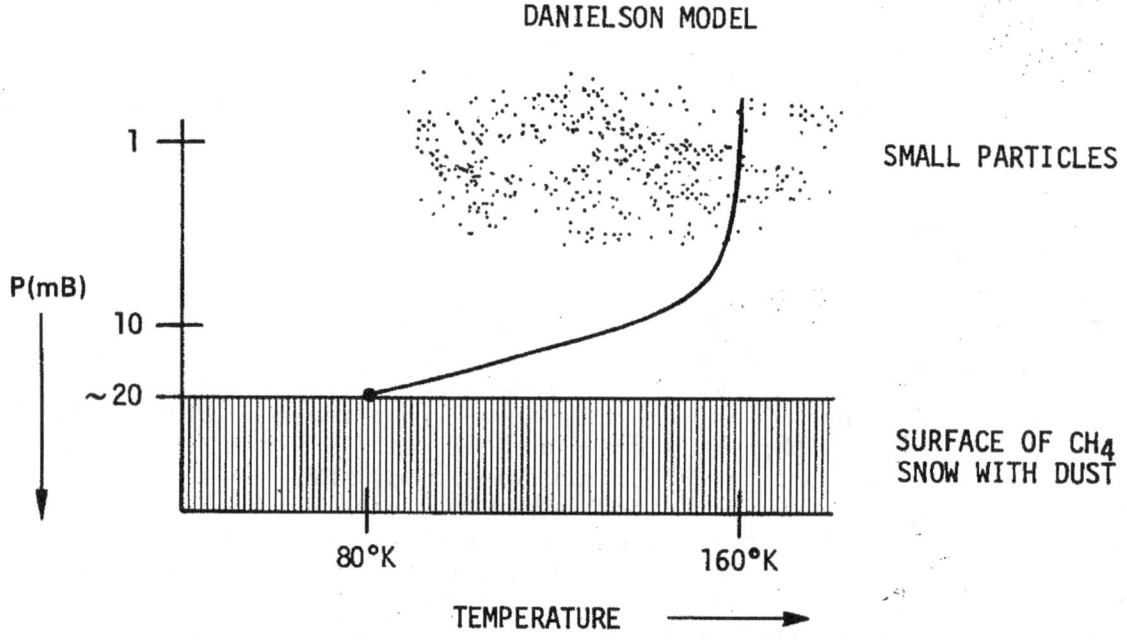

Figure 1-1. Contrasting Titan Atmospheric Models

of Titan should be close to 82°K, the value obtained for equilibrium with solar radiation when the energy is distributed over the globe by the atmosphere. Essentially this value is observed in the 20-micron region, close to the peak of the Planck curve. The question is whether the 82° temperature is that of the surface or of an elevated layer in an opaque atmosphere. The only known source of opacity above 15 μm in a cold atmosphere is pressure-induced absorption by H_2. The peak of this diffuse band is at 17 μm, and an optical depth of 2 will be produced by the observed 5 km-A of H_2 with some 15 km-A of N_2 or CH_4; the surface pressure is around 200 mb. A series of quantitative greenhouse models for H_2-CH_4 mixtures has been presented by Pollack (1973). The best agreement with the earlier broadband observations was obtained with equal abundance and a surface pressure of 440 mb. Mixtures including He were also considered, but are hard to accept because of the rapid escape of helium that must occur.

The suggestion of a warm stratosphere was based on the observed low ultraviolet albedo, which implies the deposition of the corresponding solar energy at high altitudes. Since the absorbing particles are probably too small to be efficient infrared radiators, they transfer the heat to the gas, which then radiates in any available vibration-rotation bands. The best candidates are the fundamentals of CH_4 (7.7 μm), C_2H_6 (12.2 μm), and C_2H_2 (13.7 μm). The first two are probably present (Gillett et al., 1973), and data are lacking for the third. A heuristic model by Danielson et al. (1973) puts the surface temperature at 80°K; but the cold radiating layer could probably be at some height in an atmosphere for a greenhouse model.

The presence or absence of a greenhouse due to H_2 can be decided by the shape of the spectrum around 17 μm: a negative temperature gradient implies a minimum at the center of the band. Confirmation and extension of the band structure at shorter infrared wavelengths will also be highly revealing. All these measurements appear to be technically feasible today, and many questions should be answered within a year.

Atmospheric Escape and Recycling

Several authors have touched on the problem of retention of H_2 by such a small body as Titan, and a detailed discussion has been given by Hunten (1973a). An atmosphere of pure H_2 is simply unbound, and would fly away in a few hours. If a heavier gas is present, it retards this loss, and a situation of "limiting flux" is closely approached. The composition is independent of height, except in an outer corona of H_2 that matches the limiting flux to the escape level. The flux depends only on the mixing ratio; it is about 2×10^{11} cm^{-2} sec^{-1} for 10% H_2, and 10^{12} cm^{-2} sec^{-1} for equal parts of H_2 with CH_4 or N_2. The 1/e time constant for H_2 escape is increased to around a million years. Thus, even though transient stability is assured, a source equal to the escape flux is required. A source of even 10^{11} molecules cm^{-2} sec^{-1} is difficult to find, and no really credible one has been suggested. Photolysis of NH_3 could suffice if the surface were as warm as 145°K, and if ultraviolet radiation could penetrate deep enough. The low ultraviolet albedo does not encourage this idea. Sources in the interior remain entirely speculative.

The same principles apply to helium, with only a slight change in the numerical coefficient. Radioactive decay would be expected to produce a source of around 10^5 cm^{-2} sec^{-1}, by comparison with Earth. The corresponding mixing ratio, with whatever heavy gas may be present, is about 10^{-7}.

McDonough and Brice (1973a, b) have pointed out that gas molecules lost from Titan do not escape completely, but go into orbit around Saturn. Depending on the rate of loss from the resulting toroid of H_2, there could be a significant recycling back to Titan, and a significant reduction in the net loss rate. Titan's first response to recycling would be to build up the coronal density to retain the same net loss rate as before. Thus, rather large densities may be required for the toroid; one rough estimate by Hunten gives 10^{10} cm^{-3}. If such densities are permitted, recycling would greatly ease the problem of finding an adequate source for H_2 for Titan's atmosphere. The toroid is a fascinating object in itself, and will doubtless be discussed in detail in the next few years.

Chemistry

The photochemistry of the atmosphere has many ramifications, some of which have been touched on above. Generally speaking, irradiation of CH_4 is sure to produce more complicated organic compounds and free H_2. If NH_3 is present as well, the possibilities are even greater. It is not clear that H_2 is produced fast enough by such processes to explain its probable large abundance, but some production is a certainty. At least a small production of N_2 is likely as well. Sagan and Khare have reported the production of a brown polymer in laboratory irradiations.

On Titan, most compounds with more than 2 carbon atoms will condense, many of them permanently on the surface. (Or they may dissolve in an ocean of liquid methane, if that unlikely medium is present.) The observed presence of a fine, dark aerosol to high altitudes agrees well with this expectation. It can hardly be said that life should be expected under such circumstances; but "prebiotic molecules", which are almost as interesting, should be abundant.

An interior dominated by hot ammonium hydroxide solution should also be kept in mind as a possible medium for chemistry. Radiolysis could be a source of gases, especially H_2 and N_2, that could reach the atmosphere.

Chapter 2

WORKSHOP PRESENTATIONS AND DISCUSSIONS

2.1   RADIUS AND MASS OF TITAN

D. Morrison

Introduction

Knowledge of the radius and mass of a satellite is fundamental to an understanding of its physical nature. Both quantities are needed to derive the density and the surface gravity, and the radius is also required for the interpretation of visible and infrared photometry in terms of albedo and brightness temperature. Of the two parameters, Titan's mass is by far the better known.

Titan's Mass

Titan's orbit is in resonance with that of Hyperion, and from its perturbations on the smaller satellite several investigators (e.g., Eichelberger 1911; Woltjer 1928; Jefferys 1954) have determined that its mass is 1.37×10^{26}g. These values have recently been reviewed by Kovalevsky (1970). Both from internally estimated errors and from the agreement among different authors using different techniques of analysis, it is apparent that the uncertainty in this mass is only about 1%. Even if the uncertainty were several times this value, the mass would still be one of the best-determined properties of this satellite.

Titan's Radius

The radius of Titan is, in contrast, a real problem. As seen from Earth, the apparent diameter of this satellite is only about 0.8 arcsec, and even under the best observing conditions, it is extremely difficult to measure the size of such a small object. The visual measurements have been reviewed and summarized by Dollfus (1970), and a recent note of mine (Morrison 1973) contains some additional comments on these techniques.

The most extensive and successful visual observations of the size of Titan and other satellites and asteroids of comparable angular dimensions have been made by several French observers, working primarily at the Pic du Midi Observatory with telescopes of modest aperture. The two instruments most applicable to Titan are the double-image micrometer, in which two images of the satellite are brought into tangency and their separation measured, and the diskmeter, in which an artificial image, with adjustable diameter, surface brightness, and limb darkening, is visually compared with the image of Titan. Both instruments have yielded diameters that are highly reproducible, but the systematic errors are extremely difficult to evaluate. Older, and slightly less reproducible, observations have also been made with filar micrometers at a number of observatories.

In his review of all these data, Dollfus (1970) finds a mean diameter in arcsec of 0.703 ± .10 for the filar micrometer measurements, of 0.700 ± .07 for the diskmeter measurements, and of 0.700 ± .06 for the double-image micrometer measurements. The resulting physical diameter is 4850 ± 300 km. If possible systematic uncertainties are allowed for, the actual uncertainty in this value is almost certainly larger, amounting to at least 10%.

Some calibration of possible errors is provided by recent very precise measurements of the sizes of the Galilean satellites Io and Ganymede, derived from timings of occultations of stars by the satellites. Since typical angular motions of satellites and asteroids are <0.05 arcsec/sec, a timing precision of 0.1 sec yields a precision of <0.005 arcsec in the size, which is at least an order-of-magnitude gain over the best visual measurements. The practical requirements are that the star be no more than about two magnitudes fainter than the occulting object and that the occultation be predicted sufficiently in advance to permit several photometric telescopes to be installed in the zone of visibility. Both of these requirements have been discussed in detail by O'Leary (1972). On 14 May 1971 Io occulted the 5th-magnitude star β Scorpii C. Photoelectric observations from four observing parties give a radius, on the assumption that Io is spherical, of 1830 ± 2 km (Taylor 1972). If the satellite has the figure expected for a tidally and rotationally distorted fluid, the mean radius is 1818 ± 5 km (O'Leary and van Flandern 1972). In June 1972, an occultation by Ganymede of an 8th-magnitude star was timed by three observing parties. From these data, Carlson et al. (1973) give a radius of 2635 km. The occultation radius of Io is 4% larger than the value given by Dollfus (1970) and that of Ganymede is 5% smaller; there is thus no indication of major systematic errors in the visually derived radii of these two satellites.

It is not clear whether the high accuracy of the visual measurements of Io and Ganymede will apply as well to Titan, which is smaller, has much lower surface brightness, and may, in consequence of its atmosphere, have substantially greater limb darkening. The absence of an optically thick atmosphere on any other small object makes attempts to calibrate the Titan observations against those of other satellites somewhat questionable. An alternative comparison might be with Neptune: here, the occultation diameter (Freeman and Lyngå 1970) is larger than the visual one by ∿0.07 arcsec, suggesting that limb darkening may indeed affect the visual techniques.

Stellar Occultation Measurements

I think we could all agree that the observation of an occultation of a star by Titan would be extremely valuable, both for determining an accurate radius, and, even more, for the study of its atmosphere, as discussed by Veverka (Chapter 3). If an occultation of a star of 8th magnitude or brighter were predicted, even with only a warning of a few weeks, then every effort should be made to obtain good photoelectric photometry. I note, however, that occultations even of stars of magnitude 9 or 10 could be used to measure the radius. The problem here is that our uncertainty in the positions of stars this faint precludes accurate prediction of events. In practice, an occultation will be predicted only if the star is in the SAO catalog, which is by no means complete at this magnitude level. We should seriously consider the value of a program of photographically examining the sky ahead of Titan in order to predict occultations. In order to take advantage of these predictions, it will also be necessary to have observers with high-quality portable photoelectric photometers who are willing to travel on short notice into the zone of visibility of the occultation, which will be, of course, about 5000 km wide. A modest level of funding for such a program could be a very good investment.

Summary

It is apparent from the preceding discussion that the radius of Titan is not well known at present, and that the uncertainty in the radius produces very substantial uncertainty in the density (proportional to r^{-3}) and in surface gravity, albedo, and thermal luminance (all dependent on r^2). I suggest that, for convenience, we all agree on a radius of 2500 km, which is a good round number. This value is 3% larger than the value based on the visual measurements, as interpreted by Dollfus (1970), and is 2% smaller than the value used by several recent infrared investigators (Morrison et al. 1972; Gillett et al. 1973; Joyce et al. 1973). I would also adopt an uncertainty (possibly optimistic) in this radius of ±10%. With this radius and the known mass, the density of Titan is 2.1 ± 0.6 g cm^{-3} and the acceleration of gravity at the surface is 145 ± 30 cm s^{-2}. Presumably this radius refers to approximately the cloud level, and if the clouds are a long way above the surface, the mean density of the solid body of the satellite will be higher. For many investigations, it is clear that a more accurate value of the radius would be useful. This work was supported in part by NASA grant NGL 12-001-057.

Danielson: Depending on the details of the model, the limb darkening on Titan could be either greater or smaller than on the Galilean satellites, so the sense of the correction is not necessarily to increase the radius of Titan.

Veverka: Gordon Taylor of the British Nautical Almanac Office has made a search for occultations of stars in the SAO catalog by Titan for about the next five years, and none are indicated. Thus we are unlikely to have a really high-quality event soon, although occultations of stars too faint to be in the SAO catalog but still bright enough to provide a good radius determination may take place. An alternative means of measuring the diameter of Titan is by high-speed photometry of occultations of Titan by the Moon. There will be a favorable event of this type in March 1974, and I intend to observe it.

Morrison: How accurate do you think the diameters from lunar occultations will be?

Veverka: I have not worked it out in detail, but certainly to a greater accuracy than we now have -- perhaps to a few tens of kilometers. Also, these occultations are frequent, so a number of events could be observed to refine the values.

2.2   INFRARED PHOTOMETRY AND SPECTROPHOTOMETRY OF TITAN

D. Morrison

Introduction

I would like to begin this discussion of radiometric measurements of Titan with a brief historical review. Since the discovery by Kuiper (1944) of CH$_4$ bands in the spectrum of this satellite, it has been known that Titan possesses an atmosphere, but following the discussions by Kuiper (1944, 1952), it has been assumed until recently that this atmosphere is very tenuous, with a surface pressure of only a few millibars. Such an atmosphere, as we know from the example of Mars, has only relatively minor effects on the surface temperature, at least in the hemisphere facing the Sun. Thus, it was expected that the mean temperature on the day side of Titan would be approximately that of a hemisphere in equilibrium with the insolation. At Titan's distance from the Sun, the maximum such temperature, corresponding to a blackbody facing the Sun, is 127°K. For an object with the bolometric Bond albedo of Titan, which is, according to Younkin (1973), equal to 0.27, the maximum equilibrium subsolar temperature is 116°K. The corresponding disk-averaged infrared brightness temperature would be about 0.92 times this, or 107°K. These temperatures are a few degrees lower than those that would have been computed a few years ago, primarily as a result of the higher bolometric albedo adopted here; the first authors to measure Titan's radiometric temperature in fact expected to find $110 \leq T \leq 115°$K (Low 1965; Allen and Murdock 1971; Morrison et al. 1972).

Recent observations in a number of areas have changed these expectations and have increased the significance of radiometric measurements of Titan. Most influential has been the work of Trafton (1972a, b), who discovered spectral evidence for H$_2$ on Titan and presented an important reanalysis of Kuiper's CH$_4$ observations as well as of his own. The result of this work was to suggest that the surface pressure on Titan is substantially higher, of the order of 0.1 atm. At the same time, polarimetric studies by Veverka (1973) and Zellner (1973), as well as several discussions of the low ultraviolet albedo of this satellite, showed that the atmosphere of Titan contains optically thick clouds and raised the possibility that an extensive and spectroscopically unobserved atmosphere extends to great depths below the cloud layer. An atmosphere of this magnitude might be expected to moderate variations of surface temperature so that the infrared brightness temperature of Titan would approach that of an isothermal object of its albedo and distance from the Sun, which is $T_B \simeq 84°$K.

Recent Radiometric Measurements

With this background, we now turn to the infrared observations, which have provided many surprises during the past 3 years. The first observers to note the anomalously high infrared brightness temperature of Titan were Allen and Murdock (1971), who obtained a temperature of 125 ± 2°K in a band from 10 to 14 µm (nominally 12.4 µm). Since they had expected a maximum brightness temperature of ∿115°K, Allen and Murdock concluded that there was likely to be a greenhouse effect on Titan that increased the observed 12.4-micron flux by a factor of about 2 over the equilibrium level. An earlier brightness temperature of 132 ± 5°K in the 8- to 14-micron band published without comment by Low (1965) was consistent with this suggestion.

The actual magnitude of the change in brightness temperature with wavelength, and hence of the postulated greenhouse effect, became apparent with the publication of additional observations in the infrared. Morrison et al. (1972), observing in a broad band from 16 to 28 μm (nominally 20 μm) found a temperature of 93 ± 2°K, almost as low as the equilibrium temperature for an isothermal satellite. They further documented the range of observed temperatures by quoting unpublished observations by Gillett and Forrest that yielded a disk temperature of 134 ± 2°K at 11 μm and 144 ± 3°K at 8.4 μm. They assumed that the temperature in the atmosphere increases with depth and concluded that the only abundant gas that could provide the indicated great opacity in the 20-micron band together with an opacity decreasing from 12 to 8 μm is H_2 at sufficient pressure to induce translational-rotational transitions. Further, since the bulk of the radiation from such a cool body is at wavelengths between about 15 and 50 μm, they noted that this large opacity in the 20-micron band would result naturally in a greenhouse effect. They therefore argued for a massive hydrogen atmosphere with surface pressure of hundreds of millibars and surface temperatures >150°K (see also Cruikshank and Morrison 1972). At the same time, calculations by Sagan and Mullen (Sagan 1973) indicated that a pure hydrogen greenhouse on Titan is capable of generating surface temperatures higher than 200°K. These arguments for a massive greenhouse effect have been developed extensively by Pollack (1973) and Sagan (1973) and have encouraged public interest in Titan as a possible abode of life.

A great improvement in the spectral resolution of radiometric measurements has now been achieved by Gillett et al. (1973), who have observed Titan at 8 wavelengths between 8 and 13 μm with a cooled filter-wheel spectrometer having a resolution $\Delta\lambda/\lambda \simeq 0.015$. These narrow-band data, together with several broad-band measurements in the 8-13 μm band, reveal substantial structure in the thermal spectrum, suggestive of line emission. Such structure was not predicted by the greenhouse models (Pollack 1973), but it is consistent with the hot atmosphere model recently proposed by Danielson et al. (1973) as an alternative to the greenhouse models.

One additional infrared measurement of Titan has been made, at 4.9 μm by Joyce et al. (1973). The main constituents suggested for the atmosphere of Titan -- CH_4, NH_3, H_2, and N_2 -- have no absorption bands at this wavelength, so that, aside from possible cloud opacity, a temperature at 4.9 μm might be expected to be characteristic of that near the surface of the satellite. Joyce et al. did not detect 4.9-micron radiation from Titan, but set an upper limit of 190°K. Their observation also restricts the 4.9-micron albedo of Titan to a value of less than 0.5.

All of these observed temperatures are summarized in Table 2-1, categorized as narrow-band and broad-band measurements.

Interpretations

The temperatures shown in Table 2-1 clearly show the large amount of information contained in the infrared observations, information that can be used to distinguish among models. Pollack and Danielson discuss these observations in detail relative to their respective models. I will simply make a few qualitative remarks here about the observations. The most striking feature of these data is the increase in temperature from 20 μm to 8 μm. This

Table 2-1a. Titan Narrow-Band Measurements (Gillett et al. 1973)

λ (μm)	T_b (°K)
8.0	158 ± 4
9.0	130 ± 6
10.0	124 ± 3
11.0	123 ± 3
12.0	139 ± 2
12.5	129 ± 2
13.0	128 ± 2

Table 2-1b. Titan Broad-Band Measurements

λ (μm)	$\Delta\lambda$ (μm)	T_b (°K)	AUTHORS
4.9	0.8	<190	Joyce et al. (1973)
8.4	0.8	146 ± 5	Gillett et al. (1973)
10.0	5.0	132 ± 5	Low (1965)
11.0	2.0	134 ± 2	Gillett et al. (1973)
12.0	2.0	132 ± 1	Gillett et al. (1973)
12.4	4.0	125 ± 2	Allen & Murdock (1971)
20.0	7.0	93 ± 2	Morrison et al. (1972)

trend can be understood in terms of either of two very different temperature regimes. The first assumes, perhaps influenced by our experience with Venus and the Jovian planets, that the temperature increases with depth. In this case, the 20-micron temperature corresponds to a level high in the atmosphere, and the shorter-wavelength values to lower levels. Then the opacity must be much greater in the 20-micron band, and one is led naturally to pressure-induced H_2 absorption, a high surface pressure, and a large greenhouse effect. The second model, as suggested by Danielson et al. (1973), assumes that there is a temperature inversion in the atmosphere. In that case, one seeks opacity sources in the 10-micron band, where the hot atmosphere is radiating, while the 20-micron temperature is more likely that of the surface. Such a model does not require either high temperatures or high pressures at the surface of Titan.

Perhaps the most straightforward way to distinguish between these models would be to measure the surface temperature from microwave observations, where the opacity of even a massive atmosphere and clouds is expected to be very small. Such observations have been attempted recently by Briggs and Drake at Cornell with the NRAO interferometer at a wavelength of 3.75 cm. I have been told by Briggs that they have not succeeded in separating the flux density of Titan from the noise and from the much larger flux density of Saturn. Much of the difficulty, as it turns out, is due to poor ephemerides of Titan. It is rather ironic that we are prevented from measuring the surface temperature of Titan only by an inadequate knowledge of its orbit. However, I feel that this effort to observe Titan at microwave frequencies is a very promising line of research, and that it may well provide us with the best means to confirm or refute the greenhouse models in the next year or two.

Another highly promising avenue for research lies in obtaining improved spectrophotometry in the 8- to 14-micron band and in the 17- to 28-micron band. The means now exist to obtain an excellent spectrum from 8 to 14 μm with a resolution of ∿100, and it is probable that in another year it will be possible to achieve a resolution of ∿20 in the 20-micron band. In the 20-micron band, variations of brightness temperature with wavelength due to variations in the opacity of H_2 and CH_4 are predicted by Pollack's (1973) model, but not by Danielson et al. (1973). At the shorter wavelengths also, these two models predict different spectra.

I would like to note here that, no matter what the model, Titan must emit most of its thermal radiation between 15 and 50 μm and (barring an internal heat source), the brightness temperature over this spectral region must average about 84°K, the effective temperature. Thus, I would expect no really big surprises from, for instance, a broad-band measurement at 35 μm. In the 10-micron region, in contrast, only a small fraction of the total thermal energy is emitted, even at brightness temperatures of 150°K. As a consequence, these temperatures are not constrained to be near the effective temperature of Titan, and their wavelength variations can yield valuable information on the temperature structure of the atmosphere.

Summary

In summary, the wide variation in infrared brightness temperature of Titan has, together with the spectroscopic and polarimetric studies, been largely responsible for the great interest in Titan. The original explanation of these

15

temperatures was in terms of a greenhouse effect, but that interpretation is now being challenged. Further radiometric observations, both in the infrared and at microwave frequencies, appear to provide the best prospects for distinguishing among competing models within the next year or two.

Pollack: I believe I can give some further input on the prospects for additional observations. Gillett is planning to make much more extensive spectrophotometric measurements in the 8- to 13-micron region, and I agree that could help eliminate some of the ambiguities between Danielson and myself. In addition, I think it is very important to search for structure due to pressure-induced bands at longer wavelengths. In that regard, Houck and I are hoping to make use of the NASA C-141 airborne telescope to take a look in the 16- to 60-micron region.

Sagan: A few days ago I talked to Briggs and Drake about their microwave observations and asked them specifically if they could set an upper limit now to the flux density of Titan. I was told they could certainly exclude brightness temperatures of 300°K or higher, but that without a refined ephemeris, they were not sure how much lower they would be able to go.

I would also like to ask if there are any plans for high altitude observations in the 5- to 8-micron region, from aircraft, balloons, or perhaps spacecraft? It is just terribly exciting, this increase in temperature as the wavelength gets shorter and shorter, and it certainly would be interesting to get beyond the atmospheric cut-off at 8 μm.

Morrison: I don't know of any plans. Since Titan is faint and is so close to Saturn, a substantial-sized telescope and very good pointing will be required. Perhaps something could be done from the C-141 in that wavelength range. But we should remember that, even if the brightness temperature is going up, the actual infrared flux densities are dropping very fast as one gets to wavelengths short of 8 μm.

Danielson: I would think that shorter than 8 μm you would principally be examining the CH_4 in the Titan atmosphere. You would learn something about the temperature structure in the upper atmosphere, but I don't think you would be learning anything more fundamental than that.

2.3 TITAN'S SPECTRUM AND ATMOSPHERIC COMPOSITION

L. M. Trafton

Morphology of Titan's Near-Infrared Spectrum

Titan's spectrum is remarkable both for its similarity and its dissimilarity with the spectrum of Saturn. The similarity shows up in the low resolution spectra of the two objects and also at very high resolution where there is a parallel microstructure in the spectral features of both objects. At intermediate resolution, however, there are gross differences in the sense that Titan's bands appear relatively washed out. The center of the strongest bands are filled in and the wings are enhanced in strength. The first three figures illustrate these characteristics.

Figure 2-1 shows the red and near-infrared spectra of Titan, Saturn, and the Rings. The Ring spectrum shows the telluric and solar absorption features. The resolution element is about 17 Å. Titan's 7250 Å methane band is weaker than Saturn's band as is Titan's 6000 Å methane band, but less so. On the other hand, the weak 7000 Å methane band may be stronger in Titan's spectrum. There is also a noticeable widening of Titan's bands compared to Saturn's. This widening is relatively small in this spectral region. Spectra of Titan and Saturn in the 8900 Å methane complex, ratioed to the spectrum of Saturn's Rings to remove solar and telluric features, are shown in Figure 2-2. The center of Titan's 8900 Å band is markedly filled, the 8600 Å band is filled to a lesser degree, but the 8400 Å band actually is stronger in Titan's spectrum than in Saturn's spectrum. This figure also shows the parallel microstructure clearly. Many small-scale features in Titan's spectrum also appear in Saturn's. Figure 2-3 shows the spectra of Titan's and Saturn's one-micron methane complex ratioed to Saturn's Rings. The wavelength range is about 9500 Å to 1.1 micron. Again, there is a partial washing out of Titan's spectrum with respect to Saturn's as well as a parallel microstructure. The long wavelength wing is enhanced in Titan's spectrum more than the wing at shorter wavelength. The $3\nu_3$ methane band is visible on the right.

Figure 2-4 shows a better representation of the 1.1 micron spectrum for Saturn, Titan, and Uranus ratioed to the Rings. The $3\nu_3$ band clearly is absorbing much more strongly in Titan's spectrum than in Saturn's spectrum. The Q branch is visible in each of these spectra as well as the R and P branches. There is a progressive increase of absorption in this sequence. The loss of contrast presumably results from the filling in of the continuum by overlapping lines. The Q branch is shown at higher resolution (6.6 Å) in Figure 2-5. On the left, the whole feature appears with the R(0) manifold for Saturn. On the right, the central region of this feature is depicted for Saturn's South central meridian, Titan, and Saturn's equator, respectively. Note the increased absorption going toward the limb of Saturn. According to the local continuum, the equivalent width of Titan's Q branch is about 2/3 the equivalent width of that feature in Saturn's spectrum.

The Bulk of Titan's Visible Atmosphere

Observations of the R(5) manifold of the $3\nu_3$ band are shown in Figure 2-6. The first two spectra are of Saturn and the Ring. The latter reveals only a weak water absorption feature. There follow three independent observations of Titan

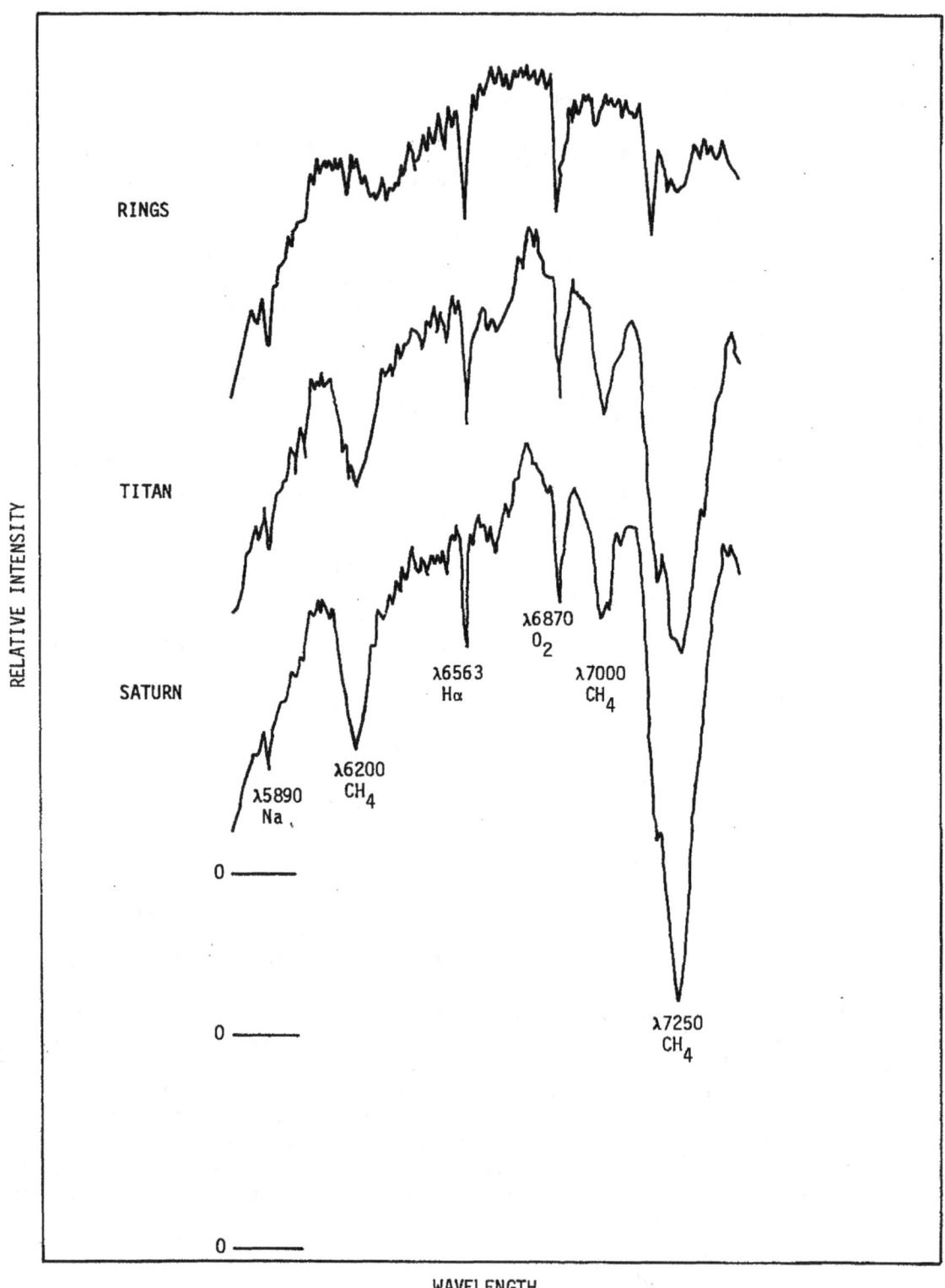

Figure 2-1. Comparative spectra of Saturn's Rings, Titan and the center
of Saturn's disc in $\lambda6200 - \lambda7250$ Å CH_4 bands. After Trafton
(1973a). Reprinted from Icarus, 21:in press, with permission
of Academic Press, Inc. All rights reserved.

18

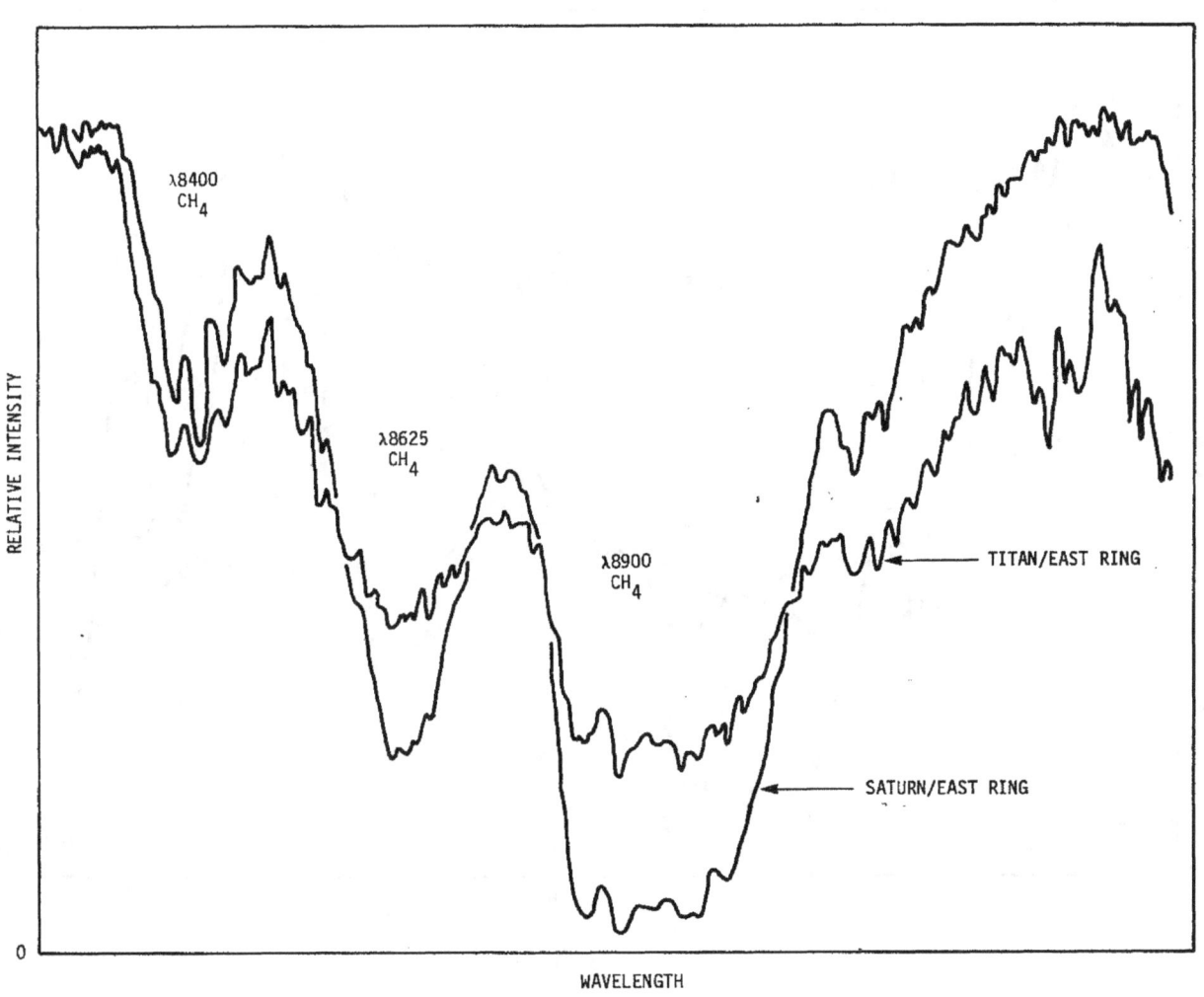

Figure 2-2. Comparative spectra of Saturn and Titan ratioed to the
Rings for the 8900 Å CH_4 complex.

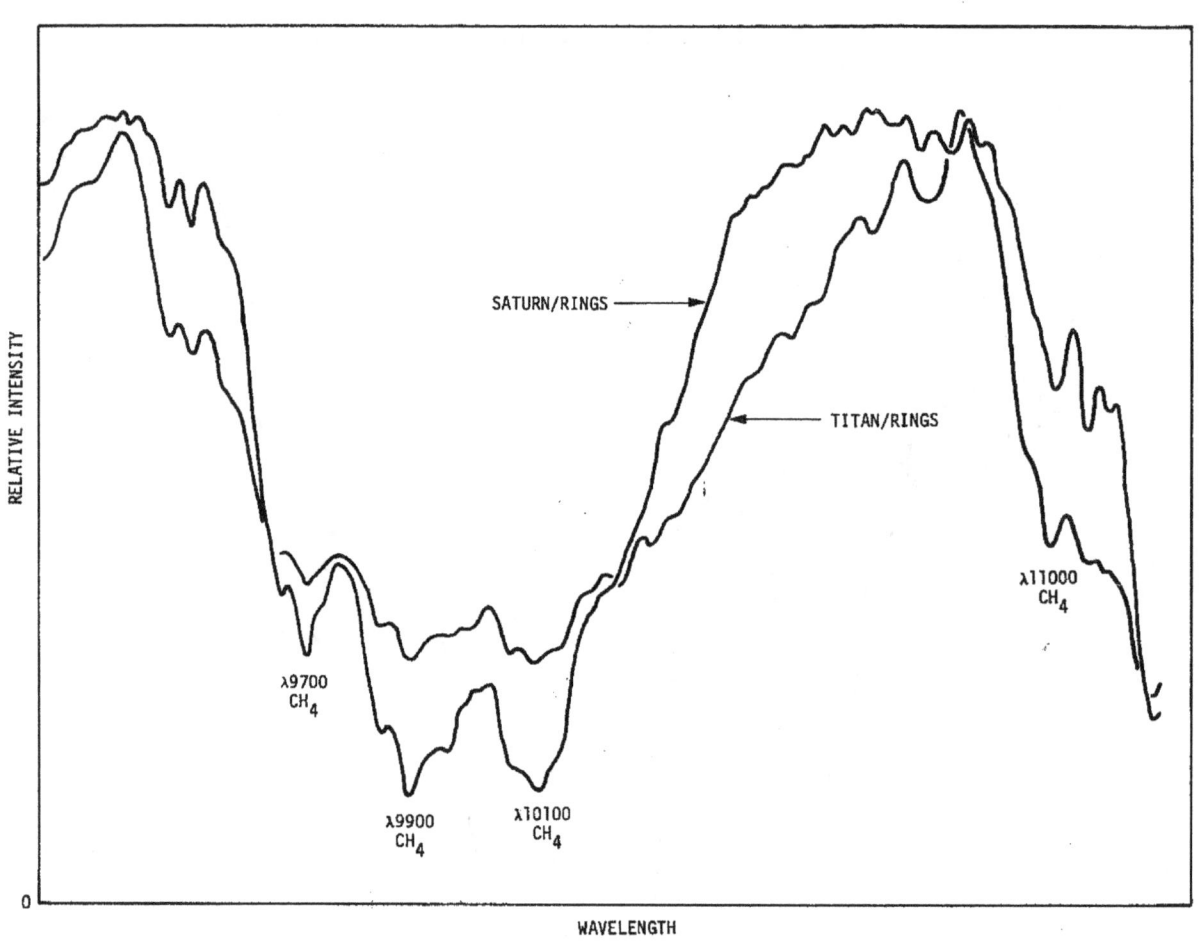

Figure 2-3. Comparative spectra of Titan and Saturn ratioed to the
Rings at the 1-micron CH_4 complex.

20

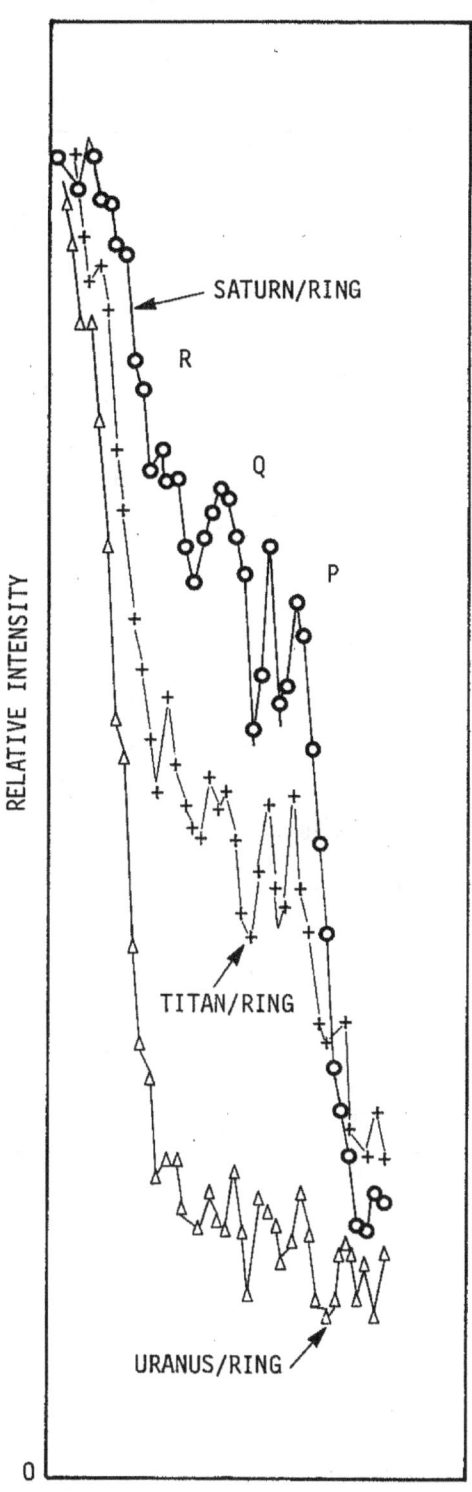

Figure 2-4. Comparative spectra of Titan, Saturn's south central meridian, and Uranus all ratioed to Saturn's Ring spectrum in the vicinity of the $3\nu_3$ CH_4 band at 1.1 μm. Note the strength of Titan's absorption. After Trafton (1973a). Reprinted from Icarus, 21:in press, with permission of Academic Press, Inc. All rights reserved.

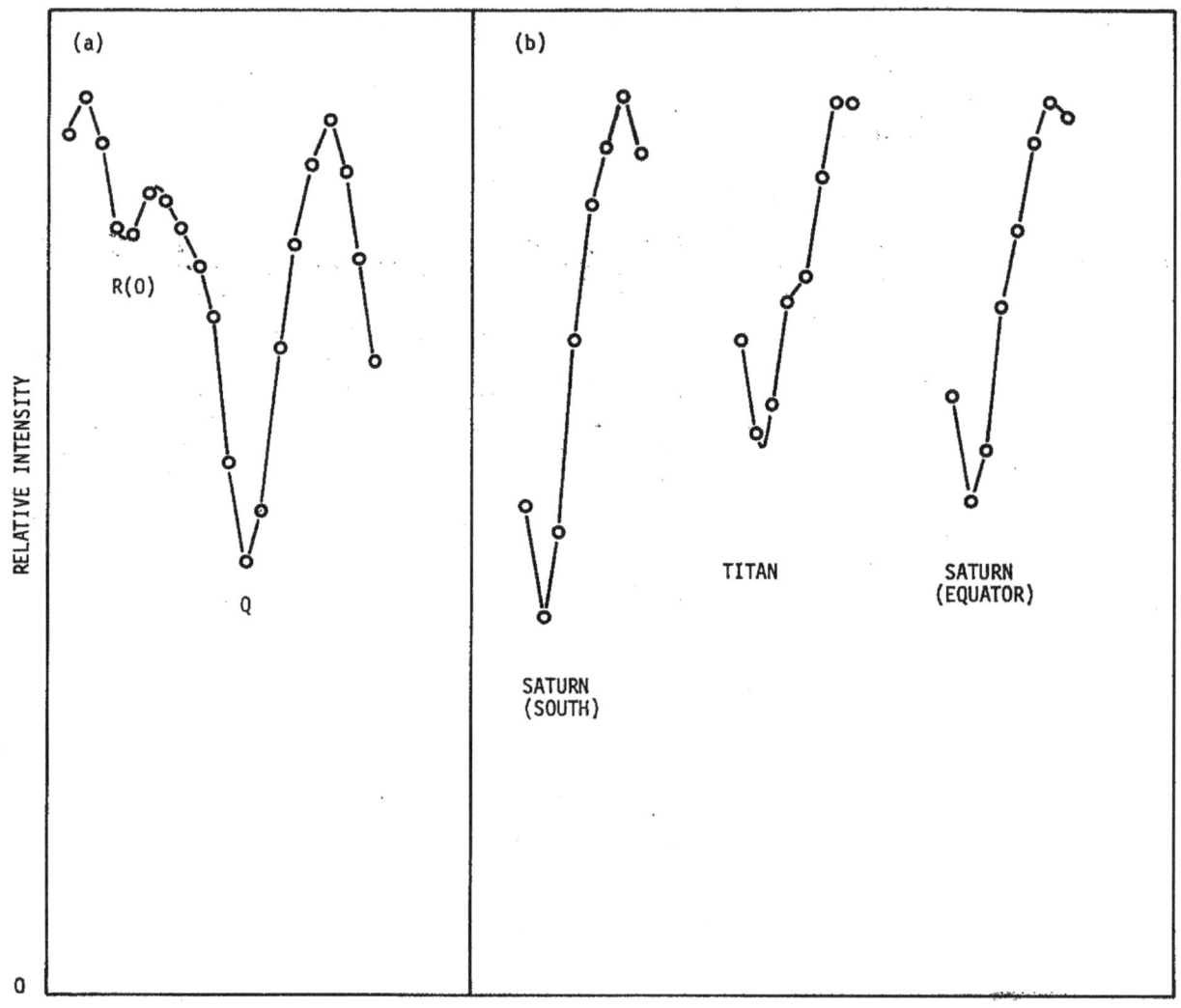

Figure 2-5. Comparative spectra of the Q branch ($\lambda 1105$) of the $3\nu_3$ CH_4 band for Titan and Saturn. (a) Saturn at low latitudes. The R(0) manifold ($\lambda 11037$) is visible here. (b) Narrow scans of the Q branch. The Saturn scans were taken with the slit set along the central meridian and excluded the Rings. After Trafton (1973a). Reprinted from Icarus, 21:in press, with permission of Academic Press, Inc. All rights reserved.

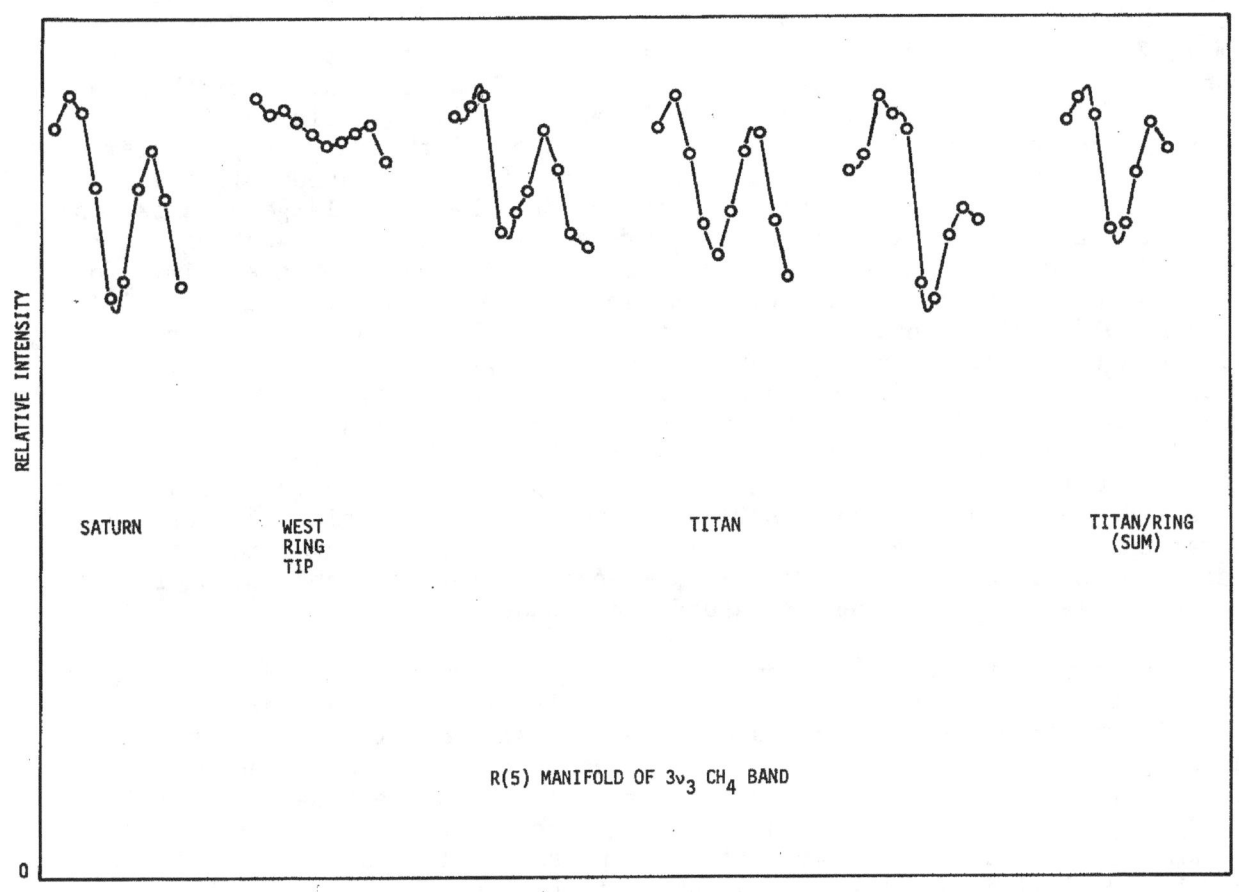

RELATIVE INTENSITY

SATURN   WEST   TITAN   TITAN/RING
                RING   (SUM)
                TIP

R(5) MANIFOLD OF $3\nu_3$ CH_4 BAND

0

WAVELENGTH

Figure 2-6.   Comparative spectra of the R(5) manifold ($\lambda 10973$) of the $3\nu_3$ CH_4
             band. The first spectrum is taken along Saturn's south central
             meridian. The ring spectrum shows the strength and position of
             telluric H_2O absorption. The next three spectra are entirely
             independent observations of Titan's R(5) manifold. The last
             spectrum is the summation of these divided by the Ring spectrum
             to eliminate the telluric absorption. The first channel of the
             Ring spectrum corresponds precisely to the first channel of the
             following Titan spectrum. After Trafton (1973a). Reprinted
             from Icarus, 21:in press, with permission of Academic Press,
             Inc. All rights reserved.

23

taken on different days; these agree fairly well. The sum of these observations ratioed to the Ring spectrum is shown on the right. The equivalent width of the manifold is just over one Ångstrom.

This result is a critical one because from it I conclude that a clear gas cannot explain the washing out of Titan's methane bands. There are four lines which make up this feature, so a lower limit for the equivalent width of one of them is 250 mÅ. Their Doppler width at this wavelength and at 90°K is about 11 mÅ. Figure 2-7 shows a curve of growth for a Lorentz line which is Doppler broadened. The ordinate is essentially the log of the equivalent width of a line in units of the Doppler width and the abscissa is essentially the line-strength abundance product in units of the Doppler width. A purely Doppler profile would qualitatively explain the washing out of the bands of Titan. Features in the square-root regime, however, would have too large a variation of the equivalent width with mean line strength to explain the washing out of Titan's bands. To establish Titan's regime, note that the lower limit for the equivalent width to Doppler width ratio for one of the lines of Titan's R(5) manifold is 23. This corresponds to 1.36 on the ordinate of Figure 2-7, well above the Doppler limit for any plausible range of conditions in Titan's atmosphere, so Doppler effects cannot wash out these spectral features. This should also be true for Titan's stronger bands since the line density appears not to exceed 7 times that for the ν_3 methane band. In this region of the diagram, the Lorentz domain, the absorption is given essentially by the pressure abundance product. Saturn's R(5) manifold is on the point of incipient saturation. Thus, throughout the entire range of physical conditions encompassing those of Saturn's and Titan's atmospheres, the line absorption is given essentially by the pressure-abundance product.

The fact that the absorption in Titan's spectrum is similar to the absorption in Saturn's spectrum indicates that Titan's smaller atmospheric pressure must be compensated by greater gaseous abundances. This is the fundamental reasoning for my upward revision by more than an order of magnitude of the amount of gas in Titan's visible atmosphere. The column abundance is at least 2 km-A of gas or 25 times that for Mars. A pure methane atmosphere corresponds to 2 km-A of gas at 10 millibars effective pressure (the "surface" pressure would be 20 millibars). Because the absorption fixes the pressure-abundance product, one could equally well explain the observation with less methane by adding another gas. For example, if there were only 100 meter-A of methane on Titan and the mean molecular weight of the atmosphere were 16, then there would have to be about 20,000 meter-A of some unknown gaseous constituent.

For 2 km-A of gas above Titan's clouds, the ratio of full width at half maximum to Doppler width (d) is 0.6. The curve of growth for this value has a slight inflection in it which could contribute to washing out spectral features only for a small range of line strengths. The value of the equivalent width at this inflection is still likely to be too low for the strong methane bands, but even for those bands which might be situated in this domain, the washing out from this slight inflection will be small. This picture of Titan's band absorption is thus self-consistent.

Danielson: On the question of the abundance, the total width of the band - never mind the washing out - does tell you something about abundances, does it not? Presumably the lines of methane that contribute to the outer wings of the band are weaker because it takes more methane to excite them. Have you ever tried to set any limits on that, or is that where you get your 2 km-A?

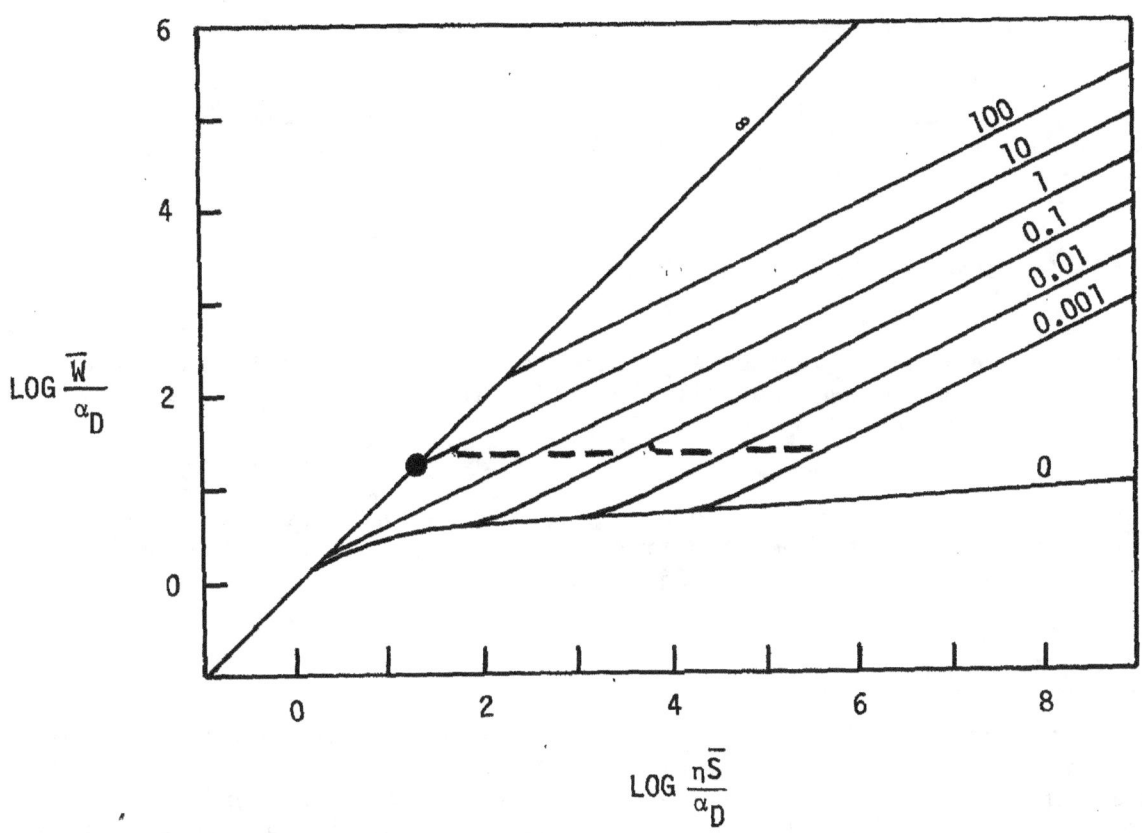

Figure 2-7. The curves of growth for a Lorentz line which is Doppler broadened. $\overline{W}$ is the equivalent width, α_D is the Doppler width, η is the air-mass factor and $\overline{S}$ is the mean line strength. The parameters represent the ratio of the full width of half maximum to the Doppler width (d). The dot denotes a typical line of Saturn's R(5) CH_4 manifold; the dashed line is the lower limit to the equivalent width of Titan's R(5) manifold.

Trafton: No. Temperature will also affect the band widths so that existing laboratory data, which are obtained at room temperature, would not be useful for abundance estimates. Comparison with Saturn's CH_4 bands is complicated by possibly different regimes in the curve of growth for the outer wings.

Pollack: Ames is now measuring the absorption of methane at different temperatures. What spectrum did you use to get your 2 km-A?

Trafton: In my original paper, I used the Q branch of the $3\nu_3$ band, but now I use the less ambiguous R(5) manifold. The results essentially agree.

Pollack: Do you mean to say you know the strength of the given line and then you go through the theoretical calculation to get its equivalent width? Is that how you do it?

Trafton: In essence, this is what I did.

Danielson: Have you not already emphasized that the continuum in all these planets is so poorly known that all kinds of things could happen just due to the continuum absorption?

Trafton: I think we do see the continuum in the vicinity of these bands I am talking about. The continuum geometric albedo is very flat in the range from 7500 Å to 10,000 Å and is about 0.36 to 0.40.

Spectroscopic Evidence for H_2

Spectroscopic evidence that H_2 is likely to be a major constituent of Titan's atmosphere is shown in Figures 2-8 through 2-11. They show spectra of Titan in the region of the (3-0) overtone of H_2. Figures 2-8 and 2-9 show the S(0) and S(1) features for the 1970 apparition, with a slit giving a resolution element of a third of an Ångstrom. On the left of Figure 2-8, a summation of 3 observations is shown and on the right a fourth observation of inferior quality owing to a larger air mass is added to these three. The arrows point to the predicted position of the hydrogen features on Titan. They differ from the position of Saturn because of the Doppler shift arising from the orbital motion of Titan. The disturbance in the continuum, which shows up in both these spectra, agrees with the predicted position within the resolution element of the slit.

Figures 2-10 and 2-11 show the S(0) and S(1) lines obtained during the 1972 apparition using a new experimental setup. We used a more sensitive photomultiplier tube having a GaAs photocathode and an echelle instead of the usual grating. The resolution element was reduced to a quarter of an Ångstrom. These data depict single observations rather than summations of a number of separate observations. Finally, the smoothing technique was a less subjective one, optimized from the modeling of the power spectra.

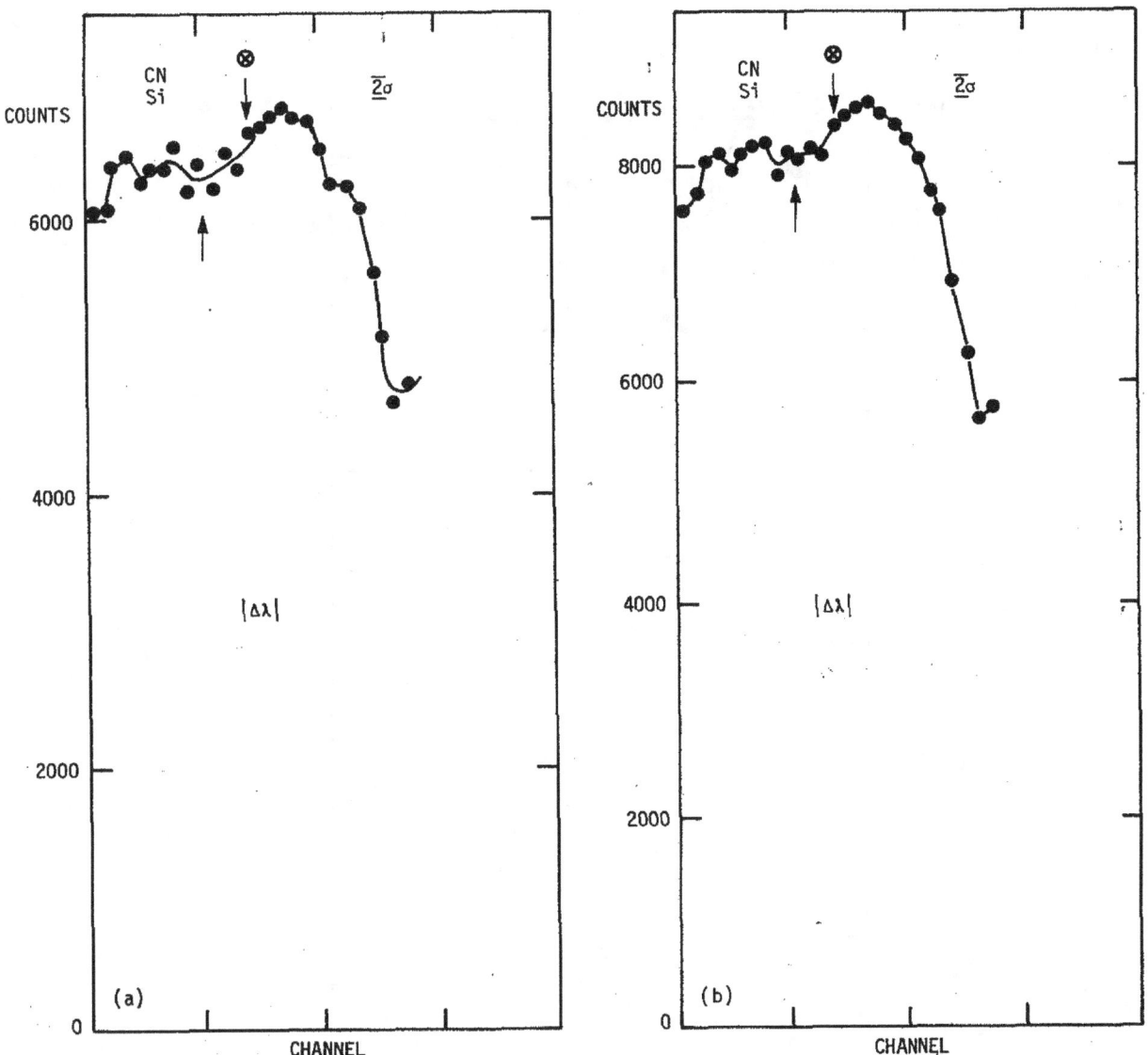

Figure 2-8. (a) Summation of the three Titan observations having the least
air mass and covering the spectral neighborhood of the 3-0 S(1)
H_2 line. The scans are superposed in Titan's reference frame,
and the dispersed positions of weak solar CN and Si lines (4 mÅ)
and a weak telluric H_2O line are indicated. The spectral resolution $\Delta\lambda$ and twice the expected standard deviation 2σ are marked.
The lower arrow points to the wavelength where S(1) absorption
would occur. (b) Similar to (a) except that all four observations are summed. After Trafton (1972b). Reprinted from The
Astrophys. J., 175:288, with permission of The University of
Chicago Press. © 1972. The American Astronomical Society.
All rights reserved. Printed in U.S.A.

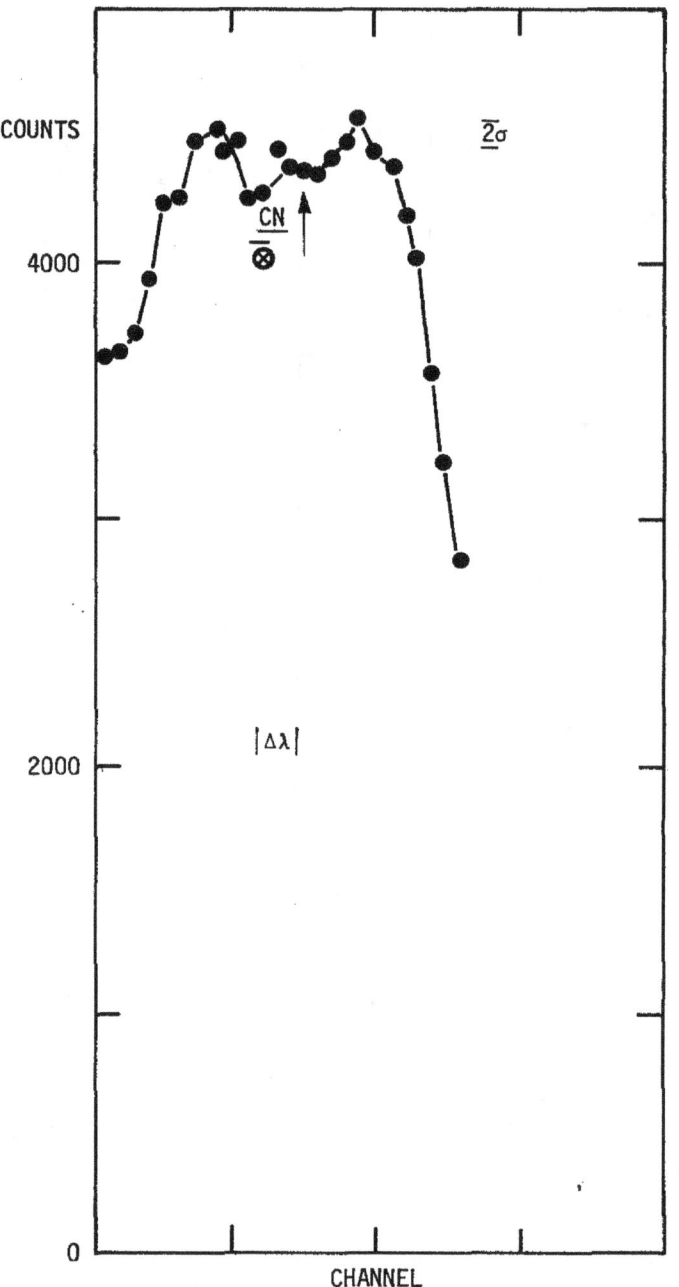

Figure 2-9. Summation of three Titan observations covering the spectral neighborhood of the 3-0 S(0) H_2 line. The scans are superposed in
Titan's reference frame, and the dashes indicate the dispersed
positions of the weak solar CN and telluric H_2O lines. The
spectral resolution $\Delta\lambda$ and twice the expected standard deviation 2σ are marked. The arrow points to the wavelength where
S(0) absorption would occur. The increased strength of the
weak telluric H_2O feature results from the higher mean air mass
and absolute humidity. After Trafton (1972b). Reprinted
from The Astrophys. J., 175:288, with permission of The
University of Chicago Press. © 1972. The American Astronomical Society. All rights reserved. Printed in U.S.A.

28

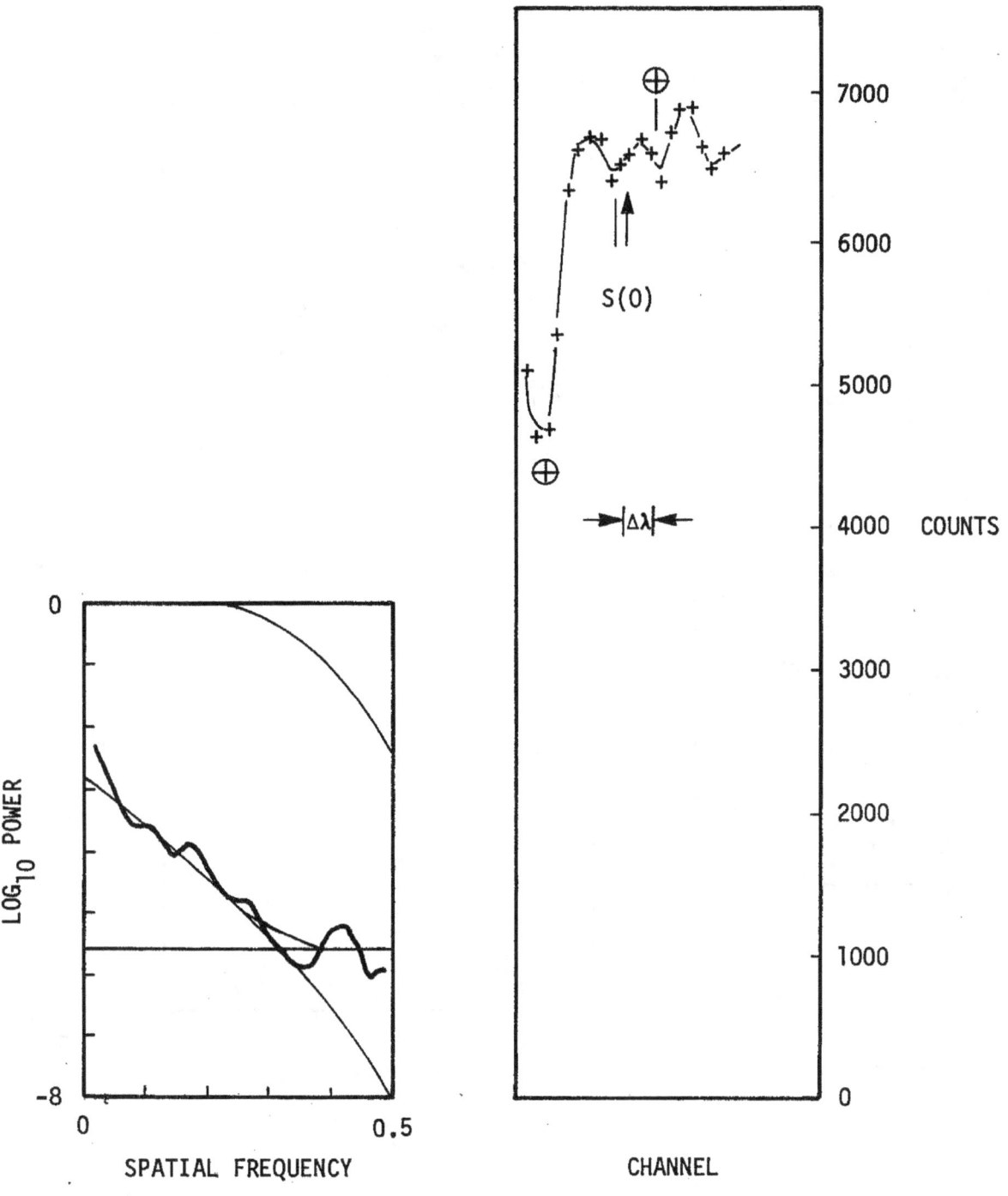

Figure 2-10. The 3-0 S(0) H_2 line for Titan during the 1972 apparition.

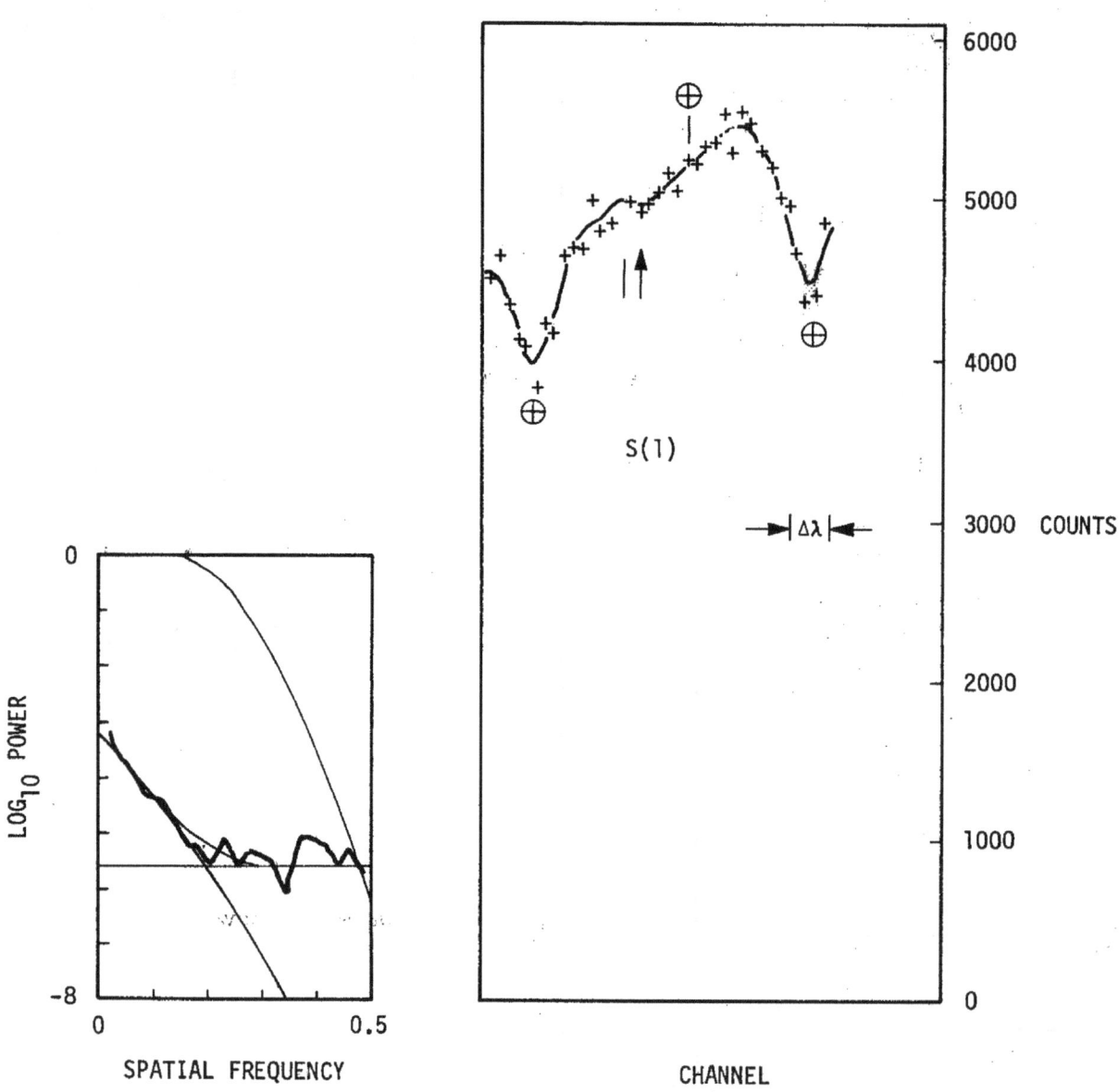

Figure 2-11.   The 3-0 S(1) H$_2$ line for Titan during the 1972 apparition.

The arrows in Figures 2-10 and 2-11 show the predicted wavelengths and the lines show the positions of the H_2 absorption features in Saturn's spectrum. The S(1) line exhibits a noisier spectrum as one can also see from the white noise level in the power spectrum. Because of the increased noise, the smoothing is heavier than for the S(0) line. Nevertheless, there is an absorption feature at the position of the arrow. The location is in better agreement with the predicted position, which is more Doppler shifted than for the S(0) line. The absorptions appear to be quite real. These two line features suggest a hydrogen abundance of about 5 km-A, assuming that they indeed arise from hydrogen. Their tentative identification rests primarily on the coincidences in wavelength for the two features.

Spectroscopic Evidence for Another Gas

Additional spectroscopic evidence of Titan's atmospheric composition exists in the anomalous enhancement of the absorption in the long wavelength wing of the 1-micron methane band. Figure 2-12 shows distinct differences between Saturn's spectrum and Titan's spectrum at resolution 6.6 Å. The most prominent difference occurs for the feature at 1.053 μm. It shows up very clearly in Uranus' spectrum.

One can find other Titanian absorption features which also show up in Uranus' spectrum and which are either undetectable in Saturn's spectrum or are only marginally detectable. The 1.05 to 1.07 μm region of the spectra of these planets is shown with resolution element 4.4 Å in Figure 2-13. This figure shows quite a number of features in the spectrum of Titan which are confirmed in the spectrum of Uranus but which are not visible in the spectrum of Saturn. As far as such features in Titan's spectrum are concerned, Titan's atmosphere resembles Uranus' a lot more than it resembles Saturn's. It supports the concept of a deep atmosphere for the planet.

Acetylene can be excluded as the source of these features. Whether they arise from other light hydrocarbons such as ethylene or ethane remains unanswered because there is almost nothing in the literature on the spectra of these molecules between 1 and 2 μm. Isotopic methane is a very good candidate for this absorber because isotopic absorptions are shifted to the long wavelength parts of the CH_4 band. If these features arise from the photolysis of CH_4, I think there would be difficulties, because you would have to explain its absorption in the atmosphere of Uranus, where most of the methane should be frozen out in the upper layers. Also, I cannot rule out the possibility that these features arise from very weak methane transitions which do not show up as strongly in Saturn's spectrum because the amount of methane visible in Saturn's atmosphere is less. In this event, these features by themselves would imply more than an order of magnitude more methane in Titan's visible atmosphere.

Sagan: Regarding your isotopic explanation, are the isotopic ratios you need in order to give the observed line strengths consistent with, say, terrestrial planets?

Trafton: That's a good point. I haven't calculated it.

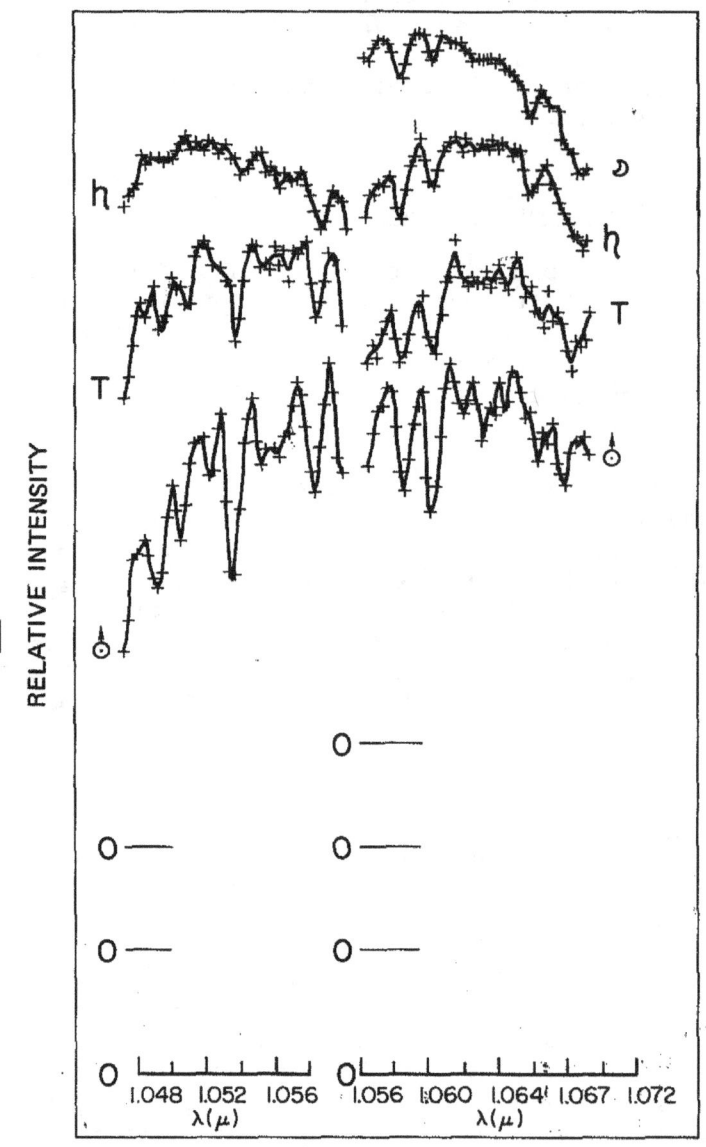

Figure 2-12. Comparative spectra of Titan, Uranus, Saturn's south central meridian, and the Moon from 1.05 μ to 1.07 μ, uncorrected for vignetting. The resolution element is 6.6 Å. The lines represent the optimally smoothed spectra and the pluses denote the data. Two observations per object were required to cover this range. Titan's spectrum includes a number of strong features which are marginally visible in Saturn's spectrum and quite pronounced in Uranus' spectrum. After Trafton (1973a). Reprinted from Icarus, 21:in press, with permission of Academic Press, Inc.

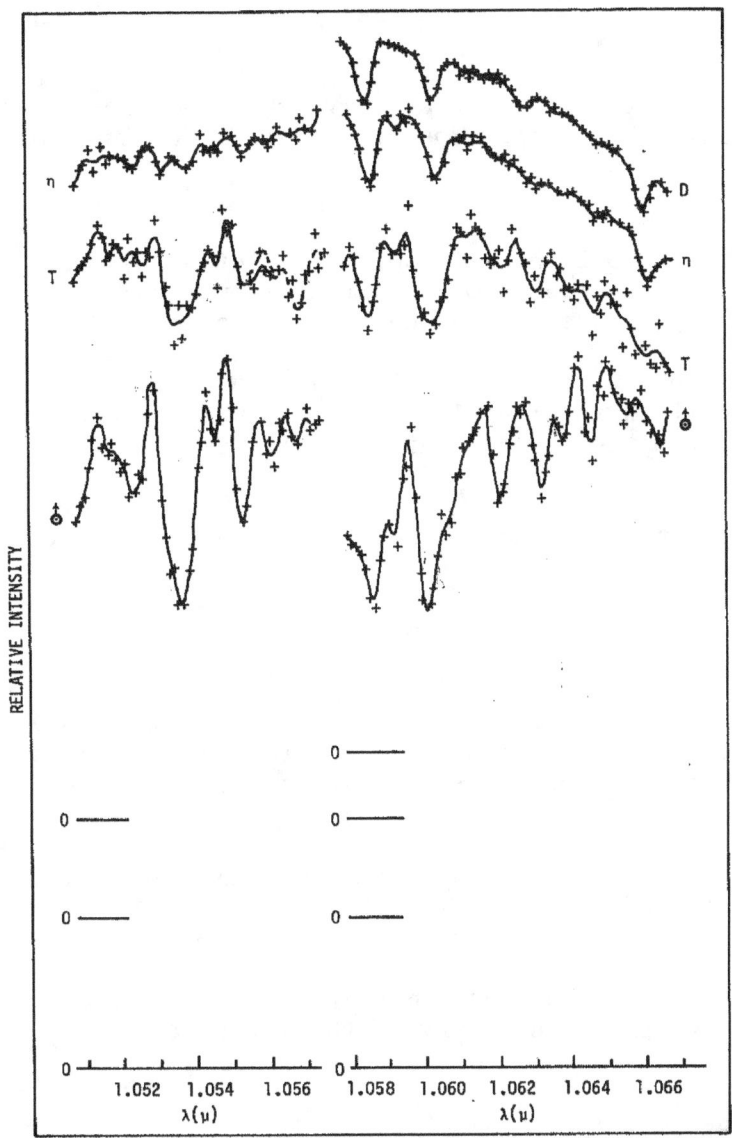

Figure 2-13. Comparative spectra for the 1.05 μm to 1.07 μm region for a 4.0 Å resolution element. These data were obtained during the latest apparition and have been normalized to the white light spectrum to remove vignetting effects. Titan's spectrum shows unequivocally the reality and strength of features hardly visible in Saturn's spectrum. The unidentified gaseous constituent responsible for these absorption features causes Titan's spectrum to approximate the spectrum of Uranus far better than the spectrum of Saturn. The visible abundance of this constituent may be much greater than that visible in Uranus' deep atmosphere owing to the probable saturation of the spectral features in Titan's spectrum and the low pressures. This points to a deep atmosphere for Titan. The unidentified feature at 1.057 μm must be of gaseous origin in order to produce this spectral variation. After Trafton (1973a). Reprinted from <u>Icarus</u>, 21:in press, with permission of Academic Press, Inc. All rights reserved.

Sagan: But you could at least see whether you're off by 2 orders of magnitude.

Hunten: The trouble with that is that very often the transition probabilities for vibration-rotation bands are grossly different for isotopic forms; again the real need is for lots of lab data.

Sagan: But surely, it would be astounding if, let's say, the C12/13 ratio of Titan were off by several orders of magnitude from what it is in comets and the Earth. I agree you have to use the right laboratory data.

Your point about it being unlikely that the absorber is a photolysis product is because it is in Uranus' atmosphere, where you have to go pretty deep in order to get optical depth unity, isn't it?

Trafton: Yes, deep in the H_2 to get optical depth unity in the CH_4.

Sagan: Have you calculated that? I mean, how does it work out quantitatively?

Trafton: No, I did not do a quantitative analysis. One really should do this accounting for the temperature profile including the effect of a temperature inversion in the upper layers of Uranus' atmosphere.

Sagan: Also, if it were a simple hydrocarbon other than methane, how would you understand that except by photolysis?

Trafton: Could there be other ways that hydrocarbons could be formed other than photolysis, perhaps in chemistry of the interior?

Spectroscopic Evidence for High-Altitude Dust

The remaining bit of spectroscopic evidence concerning Titan's atmospheric composition is the anomalous ultraviolet (UV) absorption. The relative reflectivity of Titan is very close to that of Saturn in the spectral range 3000 to 4500 Å, as is shown in Figure 2-14. This is an interesting result because the amount of gas in Titan's atmosphere should produce a marked brightening in the ultraviolet from Rayleigh scattering. Because the reflectivity is close to that of Saturn's Ring for all wavelengths in this region (the value at 3000 Å is uncertain because of strong telluric absorption), one concludes that there is no trace of Rayleigh scattering at all. This indicates that there is a strong amount of UV opacity high in Titan's atmosphere obscuring the deeper gases. The geometric albedo as a function of wavelength of Titan in this part of the spectrum is shown in Figure 2-15. The crosses depict McCord's data; he now disclaims the discrepant point at 3000 Å.

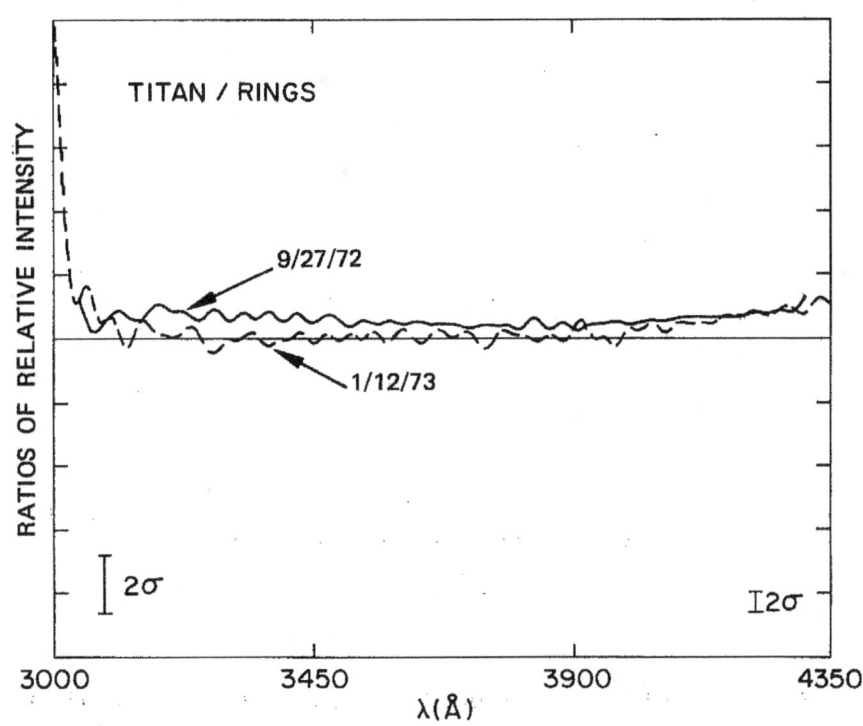

Figure 2-14. Ratio spectra of Titan vs. Rings, 9/27/72 and 1/12/73. After Barker and Trafton (1973). Reprinted from Icarus, 20:in press, with permission of Academic Press, Inc. All rights reserved.

35

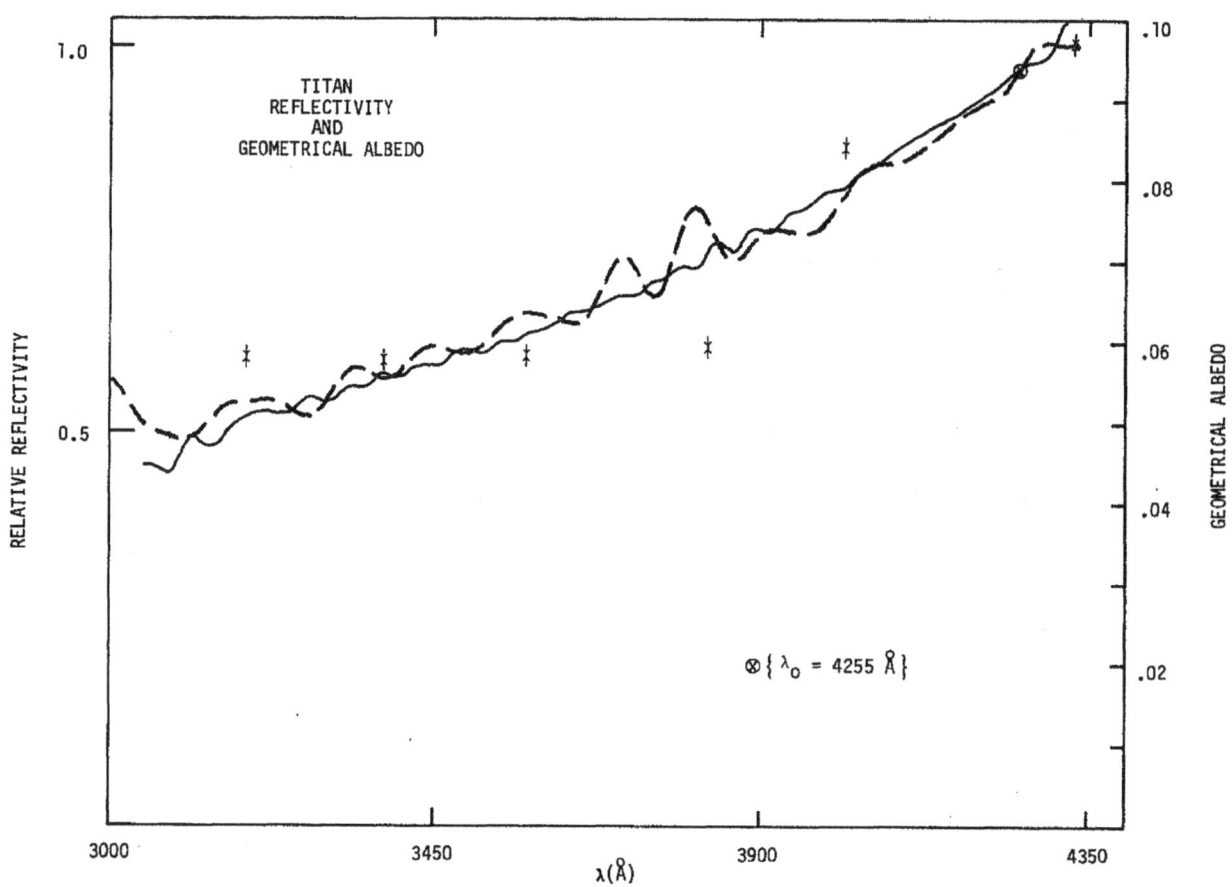

Figure 2-15. Relative reflectivity and geometrical albedo of Titan. Reflectivity normalized to 1.0 at 4330 Å. Symbols: solid curve, 9/27/72; dashed line, 1/12/73; ‡, McCord et al. (1971). Geometrical albedo of 0.094 ± 0.012 measured 4255 Å. After Barker and Trafton (1973). Reprinted from Icarus, 20:in press, with permission of Academic Press, Inc. All rights reserved.

Any homogeneous model atmosphere explaining Titan's ultraviolet albedo implies a UV absorber having a very large variation with wavelength, even more than that for Rayleigh scattering. Axel (1972) was able to explain Jupiter's anomalous UV absorption in terms of a fine dust; that is, an entity which he defines as nonscattering and varying inversely as the wavelength. This is the case for particles small compared to the wavelength of light when the complex index of refraction does not vary with wavelength. Because of the success of this model with Jupiter, an interesting question is whether one can also explain the phenomenon of Titan's UV absorption in terms of such a dust. This is possible only when there is an appropriately inhomogeneous distribution of the dust in Titan's atmosphere. If the gas density falls off more rapidly than the dust density with height, it is possible to explain the wavelength variation of the monochromatic UV albedo. I approximate this condition by a uniform dust layer overlying a layer which consists of a mixture of a fine dust and a Rayleigh scattering atmosphere. The calculations indicate that surface albedo versus wavelength is quite compatible with what one finds for the other satellites of the solar system. For a gray cloud underlying the dust layer, an opposite extreme for the underlying atmosphere, the calculations are also in fair agreement with the data. This UV opacity appears to be compatible with the observed albedo.

Veverka: What phase angles did you make these observations at?

Trafton: 4.5 degrees.

Veverka: In the ultraviolet the phase coefficient is quite appreciable and you could be off by as much as 7%.

Trafton: Then the data should be so corrected.

Morrison: And what radius did you use?

Trafton: I used the radius of 2550 kilometers.

Danielson: We have done some calculations with dust absorbing as λ^{-3} mixed uniformly in an atmosphere and, in addition, the surface albedo decreasing with wavelength and you can make the fit that way also. With that kind of model, you can fit one or two km-A of Rayleigh scattering methane.

Sagan: What is the optical depth of that model, say, at 5000 Å?

Danielson: The optical depth of the dust is about 0.2, and the Rayleigh scattering is somewhat less than that.

Sagan: Then you're seeing the surface very well.

Trafton: I used the constraint that the optical depth at 0.3 μm is at least 0.45 for Rayleigh scattering alone.   In other words, I set a lower limit to the thickness of the atmosphere.

Danielson: If dust is defined as something that absorbs as 1/λ, then it seems a very restricted definition.   Its complex index must be independent of wavelength for that to happen; otherwise, you can get anything.

Veverka: The point is, you use dust to mean anything that scatters as 1/λ the way you want it, with no particular resemblance to anything real.

Trafton: Yes.   My purpose is to see if one can explain Titan's phenomenon with the same dust theory that Axel used to explain Jupiter's.

Sagan: It seems to me the requirement for an absorbing particle high in the atmosphere is a very restrictive one in the sense that a lot of boundary conditions are being forced upon you.   For example, it seems unlikely that the source would be below because of the transparent atmosphere underneath the dust.   It looks as if your dust would have to be made up there at the top of the atmosphere and, if that is the case, it looks very much like a photochemical process.

Trafton: Yes, but it is interesting to note that the shape of Saturn's Ring spectrum in the ultraviolet is so close to that of Titan's that the same process may be operating on the particles of Saturn's Rings.

Sagan: Is it possible that all of this is solid state chemistry and doesn't involve the gas phase very much at all?

Trafton: Very possible.

Veverka: You should add Io to the list, because the spectrum of Io is very like that of Saturn's Rings.

Spectroscopic Evidence for an Elevated Cloud Layer

Turning now to an interpretation of the morphology of Titan's infrared spectrum, I indicate how these data imply a cloud layer above most of Titan's methane.   Such a cloud would most likely be frozen methane particles, in which case a temperature near 90°K would appear near its base.   Furthermore, there would be a temperature gradient at its lower boundary so that a greenhouse effect would be indicated in the lower atmosphere.   This is not necessarily in contradiction with the temperature inversion observed at high altitudes since the levels of the atmosphere in question are very different. The level of the 12-micron emission applies to optical depth unity in this feature, while optical depth unity in the visual methane bands is probably much deeper because these bands are high overtones and, hence, considerably weaker than the fundamental. Some account must be made, however, of the relative band strength and abundance of of the gas causing the 12-micron emission before the separation in these levels can be determined.

Since laboratory data for methane are not available at low temperatures for these bands, I employed Saturn's spectrum along the central meridian at elevated latitudes to derive the necessary properties of the methane bands. The absorption in this region should be similar to the spectrum of the clear gas, as absorption dominates the scattering process here. Goody's random band model applied to the 8900 Å and 1 μm complexes provided the relative absorption strengths as a function of wavelength by assuming that these bands are saturated in the sense that they lie on the square root part of the curve of growth. I have already shown above that the clear gas model of Titan's atmosphere does not work. Similarly, my attempts to fit an isotropically scattering atmosphere to Titan's bands fail, both for a finite and a semi-infinite atmosphere, since the required degree of washing out is simply not obtained. Furthermore, adding an opacity within the confines of the band, such as might result from the increased absorption of methane particles, does not explain the enhanced strength of the weaker features in Titan's spectrum, even though it washes out the structure in the bands. In order to keep the band centers from becoming too dark from the added opacity, one has to increase the volumetric scattering coefficient to reduce the scattering mean free path. This reduces the specific abundance of the gas and weakens the features in the continuum, so this model must be rejected.

The only model which I have found which works is the inhomogeneous one consisting of a high cloud layer overlying most of the methane in the atmosphere. I assume the cloud to be gray and the scattering coefficient to be constant over each band. Intuitively, this model works because the particles high in Titan's atmosphere will reflect back a fraction of the solar flux before the methane can absorb it. This will fill in all bands uniformly, but the most apparent result will be to fill in the centers of the strong bands. If the optical depth of this haze layer is not too large, the absorption in the deeper layers will be visible. By making the clear layer below the cloud layer deep enough, one can enhance the strength of the weak features to an arbitrary fraction of the strong ones. Therefore, this model appears to be satisfactory.

Pollack: As I understand it, you're saying that the scattering model fits these bands and the continuum albedo varies across the band, is that right?

Trafton: I take the "continuum albedo" constant over the band but add a gray background opacity within the confines of the band. I do not let it vary across the band although, in reality, it probably would vary by some degree if it arose from, say, absorption in solid methane. The only way to get the semi-infinite scattering model to give the required washing out is to include this background opacity, but we arrive at a contradiction by doing that since the increase in scattering required to keep the band centers from becoming too dark causes other features to be too weak.

Rasool: Could the dust be in the form of ice? Can you have liquid particles in the cloud layer? What do you mean when you day "dust" -- liquid or solid?

Trafton: When I say "dust", I am talking about the ultraviolet opacity which occurs at wavelengths shorter than 4000 Å. In the infrared, the high cloud layer no longer looks like a dust layer, but like a scattering layer. I assume that its ultraviolet opacity varies as $1/\lambda$, to see if Titan can be explained with the model Axel used for Jupiter.

Pollack: Did I understand you to say that the 1 μm observation indicated scattering as well as absorption, is that correct?

Trafton: Yes. I believe the continuum albedo at 1 μm is about 0.36 to 0.39.

Danielson: Can you summarize the key reason or reasons why the washing out of the bands cannot be explained by the fact that lines on Titan must be quite narrow? It seems to me the washing out of the bands is a very important part of the argument that you need for a scattering layer. If that's true, your explanation will do it, but it may not be unique.

Trafton: I believe you do need it because the lines are so strong their Doppler cores are essentially black. Any further absorption depends on the absorption in the Lorentz wings.

Danielson: But there may be spaces between the lines.

Trafton: However, growth in absorption comes from the Lorentz wings; this will not wash out the structure of the bands.

Danielson: But doesn't that mean they make it black between the cores?

Trafton: No. The observed lower limit to the equivalent width of Titan's R(5) manifold, a relatively weak feature, implies that the absorption in the center of its lines is black; any further absorption must therefore occur in the Lorentz wings. The curve of growth in this regime will be the square root asymptote, which has a relatively steep slope. You can't explain washing out of the contrast with a slope that steep. You could explain it with the shallow Doppler curve of growth, but I attempted to show that Titan's physical regime lies above this curve so that this situation is excluded.

Danielson: Even in that regime with certain line spacings, you can get a washing out. Make little spaces big enough between the lines and you surely get it. So, it takes detailed modeling to really establish these facts, doesn't it?

Trafton: What we are measuring, and actually accounting for in the analysis, is an average of the rapidly varying monochromatic albedo over the 17 Å resolution element of the spectrograph. Random band models, as previously shown, establish the relationship between the albedo in the band and the behavior with wavelength of the mean line strength in terms of the equivalent width of a mean line and the curve of growth.

Danielson: That's correct, but one parameter in there is a mean line spacing. This spacing must depend on the total abundance you have, because if you had much more abundance in the line of sight on Titan, you would bring in far more lines and hence the mean line spacing would change, I suspect.

Trafton: There is that possibility.

Danielson: Your models are based on the fact that the mean line spacing is the same for Saturn and Titan, and then your calculations indicate that the clear spaces between the lines cannot explain the washing out and hence you need some scattering.

Trafton: That is right. I do assume that the line spacings are the same, going from Saturn to Titan at a given wavelength. To get a washing out, one must find that the ratio of the equivalent width of a mean line to the mean line spacing decreases with respect to the same ratio when no new lines are added as the abundance is increased. Adding weak lines reduces the mean line spacing; but it also reduces the mean equivalent width. It is not at all clear that the above ratio should become significantly less.

Note: This article is, in part, a summary of publications and preprints by Barker and Trafton (1973), and Trafton (1972a, 1972b, 1973a, 1973b).

2.4 PHOTOMETRY AND POLARIMETRY

J. Veverka

Introduction

This paper is a review of currently available information on the photometry, polarimetry, and narrow-band spectrophotometry of Titan. It is convenient to divide the discussion into five major categories:

(1) Brightness and color as a function of orbital position,

(2) Brightness and color as a function of solar phase angle,

(3) Geometric and Bond albedo,

(4) Reflectance as a function of wavelength,

(5) Polarization as a function of solar phase angle.

These topics are dealt with in turn in the next five sections. The final section contains conclusions and a summary of the best, currently available data.

Brightness and Color: Orbital Position Dependence

Titan revolves about Saturn once every 16 days. Originally Pickering (1913) announced a variation in brightness of 0.24 magnitude with this period. Harris (1961), however, conclusively showed that this reported variation is spurious and is due to errors in the magnitudes assigned to comparison stars. His own measurements at McDonald showed no definite variations within ±0.08 magnitude in the V. Nevertheless, visual observers occasionally report semipermanent markings on Titan (for example, Lyot, 1953) suggesting that brightness variations may occur.

Accordingly, UBV observations of Titan were carried out on 14 nights during the 1968-69 opposition with the Harvard 16" reflector (Veverka, 1970). At this stage it was assumed that the brightness and colors do not change significantly with solar phase angle. Nine of the observations were obtained during one revolution of Titan about Saturn: from January 9 to January 17, 1969. During this time the solar phase angle changed only from 6°.1 to 6°.0. In the V, no variations in brightness, related to orbital position were found in excess of ±0.04 magnitude. To about the same degree of accuracy no variations in the (B-V) and (V-B) colors were detected.

Similar conclusions are reported by Blanco and Catalano (1971). The scatter in their V measurements is slightly smaller than that quoted above but the scatter in their (B-V) and (U-B) measurements is greater. Again, McCord, Johnson and Elias (1971) found no change in Titan's brightness with orbital phase at 0.56 μm.

A recent joint project between the University of Hawaii and Cornell University yielded high quality photometry of Titan at six wavelengths (Noland et al., 1973). Sufficient observations were obtained to permit a separation of brightness changes due to orbital position from those due to changes in the solar phase angle. No evidence for brightness variations was detected as is apparent from the data presented in Table 2-2. Similarly, these data do not show any color changes related to orbital position.

The conclusion is that if short term changes in the brightness of Titan occur, their amplitude does not exceed ±0.02 magnitude, and they are not related to Titan's orbital position. The possibility of long term (secular) changes is considered in Section 3.

If the atmosphere of Titan is optically thick, we would not expect any variations in brightness with orbital position. If the atmosphere is optically thin, then the distribution of surface brightness must be quite homogeneous, unlike that of other Saturn satellites and of the Galilean satellites of Jupiter. The absence of established color changes sets limits on the extent and contrast of the atmospheric changes reported by some visual observers.

Brightness and Color: Phase Angle Dependence

Important information about Titan can be obtained by measuring the phase coefficients at various wavelengths. At small phase angles a smooth surface or a thick Rayleigh atmosphere will have an imperceptible phase coefficient (perhaps .002 mag/deg). A microscopically rough surface with no overlying atmosphere (like that of the Moon) will have an appreciable phase coefficient (say 0.025 mag/deg).

Since the brightness of Titan is independent of its orbital position, its apparent magnitude, reduced to mean opposition distance (Harris, 1961), can be expressed as:

$$m(\alpha) = m_o + \beta \cdot \alpha$$

where: α = phase angle,

m_o = magnitude at opposition,

β = phase coefficient in units of mag/deg.

The phase coefficients for Titan from the Hawaii-Cornell photometry are listed in Table 2-3 and are shown in Figure 2-16. The only other determination of a phase coefficient is by Blanco and Catalano (1971). Although these authors fit their data to a quadratic expression in α, a linear equation yields an equally good fit with $\beta(V) = 0.006 \pm 0.001$. This value of the phase coefficient is in good agreement with the Hawaii-Cornell data.

43

Table 2-2. Orbital Brightness and Color Solar Phase Angle Dependence

WAVELENGTH	AMPLITUDE (MAG)*
u (0.35 μm)	≤ .018
v (0.41 μm)	.018
b (0.47 μm)	.006
y (0.55 μm)	.010
R' (0.63 μm)	.015
I' (0.75 μm)	.015

* Upper limits on probable amplitudes.

Table 2-3. Phase Coefficients of Titan (Noland et al., 1973)

FILTER	WAVELENGTH	PHASE COEFFICIENT (MAG/DEG)
u	0.35	0.014 ± 0.001
v	0.41	0.010 ± 0.001
b	0.47	0.006 ± 0.001
y	0.55	0.005 ± 0.001
R'	0.63	0.002 ± 0.001
I'	0.75	0.001 ± 0.001

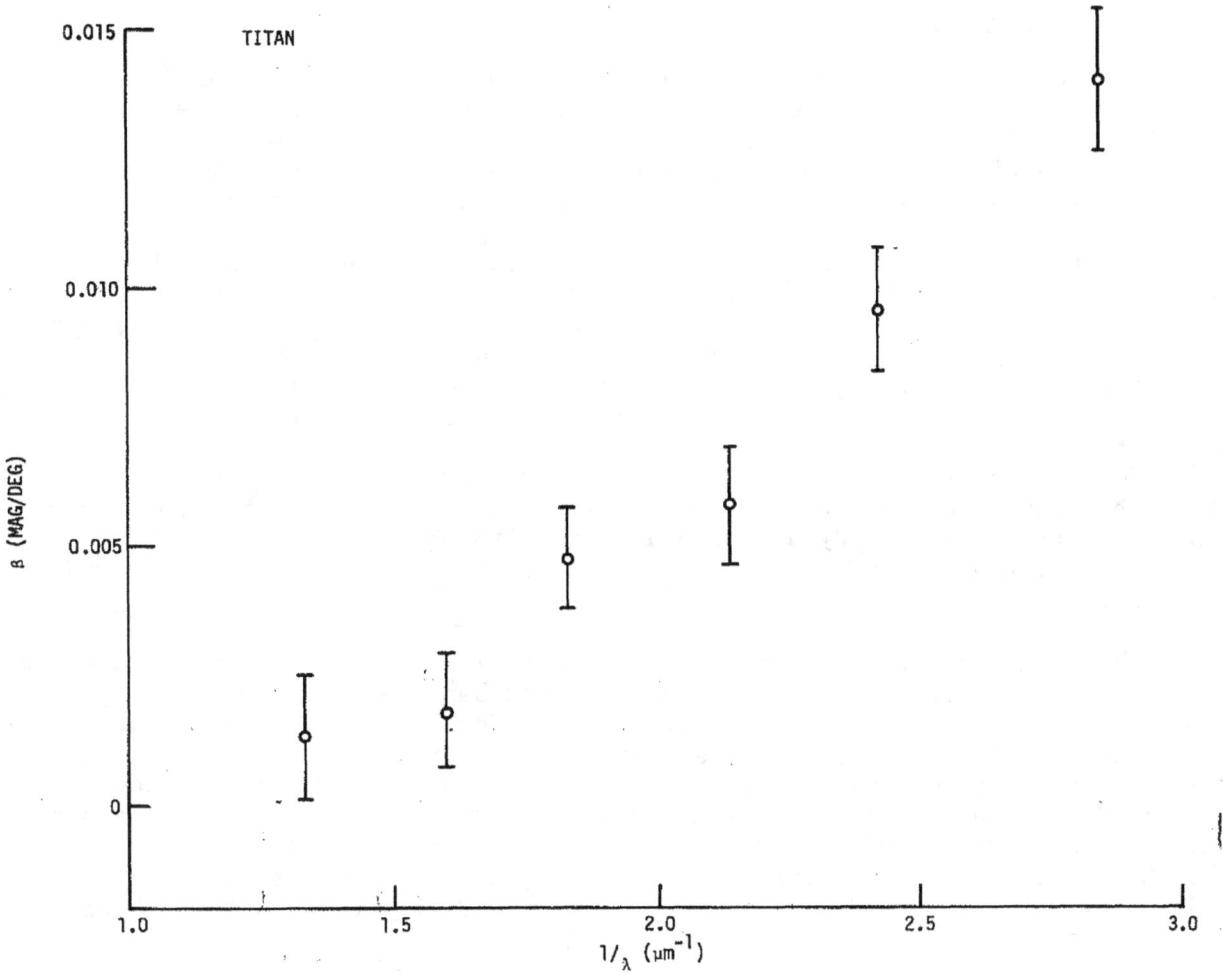

Figure 2-16. The wavelength dependence of Titan's phase coefficient. After Noland _et al._, 1973.

45

The information in Figure 2-16 can be used to set limits on possible Titan models. Optically thick Rayleigh scattering atmospheres can immediately be excluded, as can models in which most of the light comes from a solid surface (Noland et al., 1973). Furthermore limits can also be imposed on allowable cloud models (M. Noland, work in progress), using methods similar to those used by Arking and Potter (1968) in the case of Venus.

Geometric and Bond Albedo

Four determinations of the mean opposition magnitude of Titan in the V are listed in Table 2-4. The values agree to ±0.02, but a slight secular brightening between 1961 and 1971 is possible. In what follows we adopt the Harris values of the UBV colors since they agree with other determinations. Thus we have (B-V) = +1.30 and (U-B) = +0.75 reduced to mean opposition.

The corresponding values of the UBV geometric albedos given by Harris (1961) are still viable: p_U = 0.06; p_B = 0.12; p_V = 0.21. The low value of the geometric albedo in the U places a limit of $\tau(0.36 \ \mu m) \leq 0.16$ on the optical depth of any pure Rayleigh scattering atmosphere. This upper limit can be lowered considerably using the OAO-2 observations of Caldwell et al. (1973) who report p (0.26 μm) = 0.05. This translates into $\tau(0.36 \ \mu m) \leq 0.04$.

Younkin has reported (p. 154) new measurements of Titan's geometric albedo between 0.50 and 1.08 μm. The maximum value is said to be 0.37 at 0.68, 0.75 and 0.83 μm. By assuming an effective phase integral $\bar{q}$ = 1.3, consistent with a cloudy atmosphere, he estimates the bolometric Bond albedo to be 0.27.

There is a difficulty with the geometric albedo information beyond 0.6 μm which must be noted. Harris (1961) gives broad-band values of 0.32 in the R (0.69 μm) and 0.27 in the I (0.82 μm), whereas Younkin (1973) quotes narrow-band values of 0.37 near these wavelengths. That the values of Harris are lower is consistent with his use of broad filters in a spectral region of deep absorption bands. McCord et al. (1971) attempt to relate their narrow band measurements to the V measurement of Harris. If their transfer relationship were accepted, their data would imply narrow-band geometric albedos at 0.68 and 0.82 μm of about 0.27 and 0.21, respectively, in disagreement with both Harris and Younkin.

Spectral Reflectance of Titan

McCord et al. (1971) measured the spectral reflectance of Titan from 0.3 to 1.1 μm (Figure 2-17) and found it remarkably similar to the spectrum of Saturn's equatorial belt. Large methane absorption bands are present in both spectra beyond 0.6 μm, and both spectra show steep drop-offs from 0.6 to 0.4 μm. The similarity of the spectra outside the methane bands suggests that the material causing the coloration of the bands of Saturn is present on Titan as well. Below 0.4 μm McCord et al. find that the spectra differ appreciably. The ultraviolet turnup in Saturn's spectrum, probably due to a significant Rayleigh scattering component, is absent in the Titan spectrum. Recent measurements by Caldwell et al. (1973) down to 2600 Å, and by Barker and Trafton (1973) between 3000 and 4350 Å confirm that Titan is dark in the UV. More recent spectral reflectance measurements between 0.5 and 1.08 μm have been reported by Younkin (1973), but have not been published in final form.

46

Table 2-4. Mean Opposition Magnitudes of Titan

SOURCE	MEAN OPPOSITION MAGNITUDE (V)
Harris (1961)	+ 8.39
Franklin (1969)	+ 8.37
Veverka (1970)	+ 8.37
Blanco and Catalano (1971)	+ 8.35

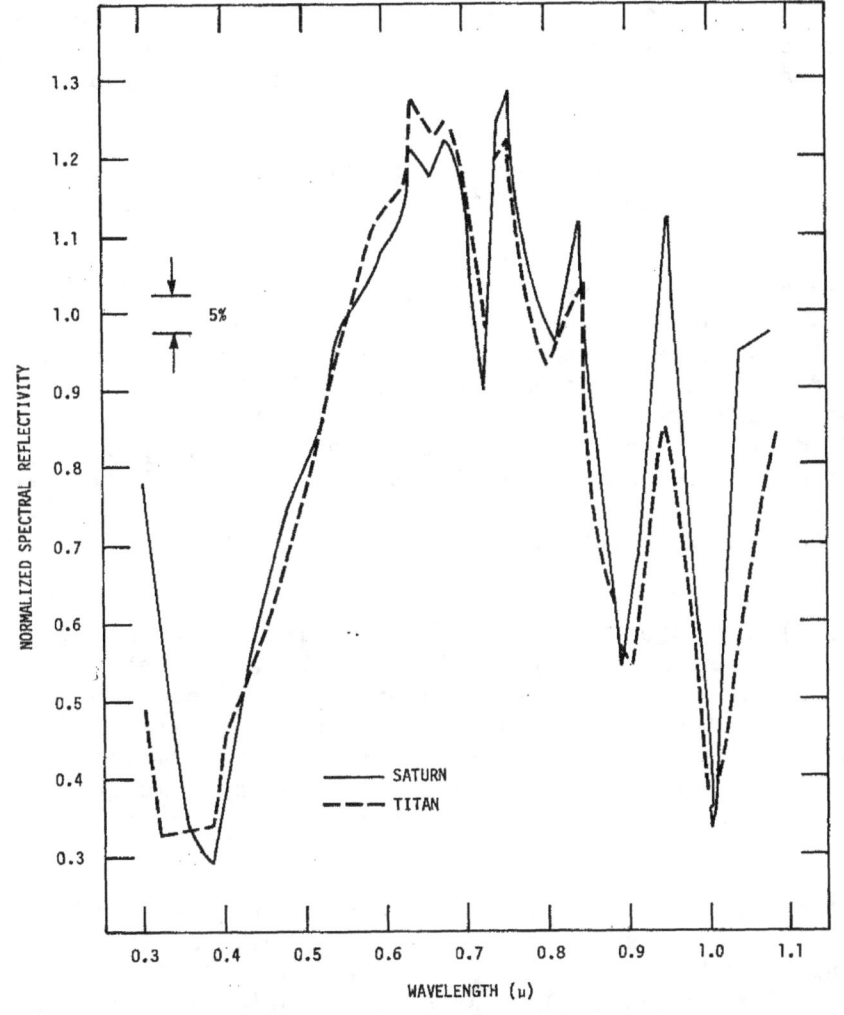

Figure 2-17. The spectral reflectance curve of Titan compared with that of Saturn's equatorial belt, scaled to unity at 0.56 μm. After McCord et al. (1971). Reprinted from The Astrophys. J., 165: 422, with permission of The University of Chicago Press. © 1971. The University of Chicago. All rights reserved. Printed in U.S.A. (The turn-up of the Titan spectrum at 0.3 μm is probably spurious.)

Polarization Phase Angle Dependence

Veverka (1970; 1973) measured the disk-integrated polarization of Titan in white light at phase angles ranging from 0°.4 to 6°.1 during the 1968-69 opposition. The observed polarization was small, but positive throughout this interval. By combining this fact with Titan's low geometric albedo in the U, the observations suggest a model in which an optically thin Rayleigh atmosphere overlies an opaque cloud deck.

The observations are shown in Figure 2-18. Note that Titan is intrinsically dark having geometric albedos of 6, 12, and 20% in the U, B, V, respectively (Section 3). Polarization curves of solid surfaces which are this dark tend to have well-defined negative branches at small phase angles, unless those surfaces are unusually smooth -- and there is no reason to expect any planetary surface to be optically smooth. Only for surfaces having very high reflectances (say greater than 50%) does the negative branch disappear and the polarization curves begin to resemble that of Titan. (The disappearance of the negative branch is related to the fact that multiple scattering within the surface achieves a dominant role.)

However such bright materials cannot explain the polarization curve of Titan, since the geometric albedo of Titan is very low: about 20% in the visible, and not 60%!

It is instructive to compare the polarization curve of Titan with those of the Moon, Mars, and Saturn (Figure 2-18). The comparison with Mars is especially interesting since Mars is similar to Titan in both color and albedo. (According to Harris, 1961, the geometric albedos for Mars are 5, 8, and 15% in the U, B, and V, respectively. The corresponding values for Titan are 6, 12, and 21%.) We know that Mars has an optically thin atmosphere, and the polarization that we see in the visible is that of the relatively dark Martian surface. The polarization curve of Titan is quite different, and in fact bears a strong resemblance to that of Saturn, a planet which certainly has an optically thick atmosphere.

Veverka (1973) discussed three a priori possible Titan models shown in Figure 2-19. Model I has an optically thin Rayleigh atmosphere (with possibly an occasional cloud) above the true surface of Titan. Since the geometric albedo of Titan is low, the surface must in this case be dark, and we should, as in the case of Mars, see negative polarization at very small phase angles (unless the surface of Titan is unusually smooth, which seems quite unlikely). Since the observed polarization is on the contrary always positive, Model I can be eliminated.

Either of the two remaining models can explain the observed polarization curve. In Model II, we have an optically thick Rayleigh atmosphere (again, with possibly occasional clouds). Calculations by Whitehill (1971) predict a disk-integrated polarization of about +0.3% at a phase angle of 6° for this case, which is compatible with the observations. However Model II is easily rejected on photometric grounds. Since the geometric albedo of Titan in the U is only 6%, the total optical thickness of any Rayleigh atmosphere cannot exceed $\tau \sim 0.16$ (Evans, 1965). In fact, the value of $p(0.25 \ \mu m) = 0.05$ found by Caldwell *et al.* (1973) reduces this upper limit to $\tau(0.36 \ \mu m) \leq 0.04$. This certainly is not optically thick and Model II must be rejected.

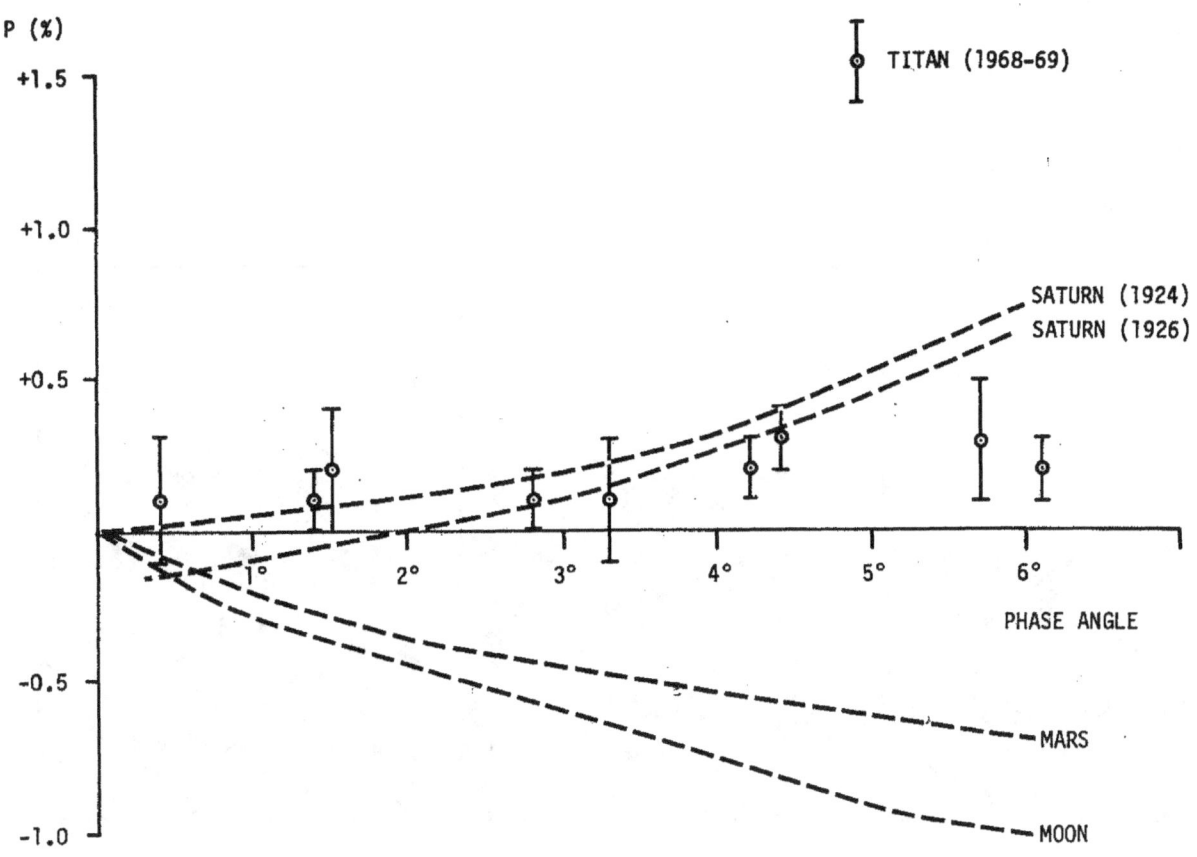

Figure 2-18. Titan measurements compared with polarization curves of the
Moon, Mars and Saturn in integrated white light (Veverka, 1973).
The Saturn measurements were made at the center of the disk
and were found to be slightly variable from year to year.
After Veverka (1973). Reprinted from Icarus, 18:659, with per-
mission of Academic Press, Inc. Copyright © 1973 by Academic
Press, Inc. All rights or reproduction in any form reserved.

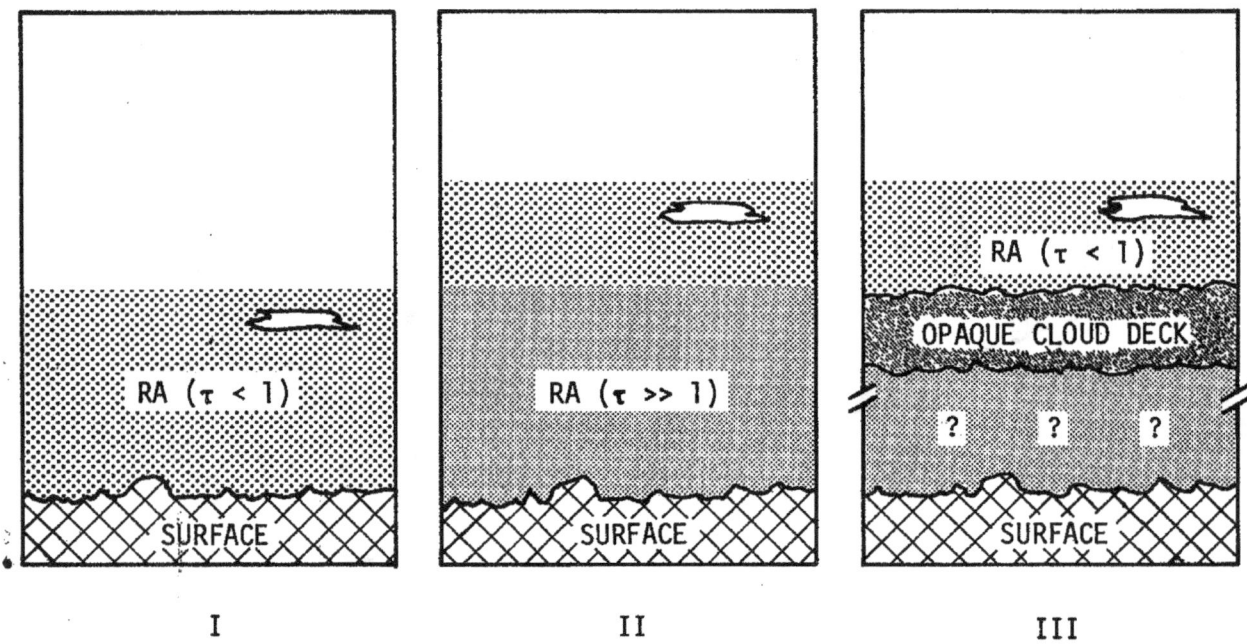

Figure 2-19. Schematic representation of three models of the Titan atmosphere: (I) optically thin Rayleigh atmosphere (RA); (II) optically thick Rayleigh atmosphere; (III) optically thick cloud deck. After Veverka (1973). Reprinted from Icarus, 18:660, with permission of Academic Press, Inc. Copyright © 1973 by Academic Press, Inc. All rights of reproduction in any form reserved.

In the last model, Model III, we have an opaque cloud top, above which there is a small amount of Rayleigh atmosphere: less than $\tau \sim 0.04$ from the discussion above, a situation similar to that which one might be expected to obtain on Saturn. (Recall that Saturn and Titan have similar polarization curves.) It is also interesting to note in this context that McCord, Johnson, and Elias (1970), found that the spectral reflectance curves of Saturn and Titan are almost identical.

Veverka (1973) concluded that both the photometric and polarimetric properties of Titan can best be explained in terms of a Saturn-like model (Model III), in which there is an opaque cloud deck overlain by an optically thin amount of Rayleigh atmosphere. The unusual red color of Titan would then be due to an absorber of blue and ultraviolet light within the cloud deck. This substance might well be the same as that in the clouds of Saturn.

New polarization measurements of Titan in three colors (0.36, 0.52, and 0.83 μm) were obtained by Zellner (1973) during the 1971-72 opposition (Figure 2-20). They confirm and considerably improve upon the earlier white light measurements by Veverka (1970; 1973). Zellner concluded that his observations "are not consistent with scattering from either an ordinary planetary surface or a pure molecular atmosphere. Apparently an opaque cloud layer with a strong UV - absorbing constituent is needed."

Zellner's conclusions are in part based on model calculations using the Rayleigh-Chandrasekhar theory for the polarization produced by a pure molecular atmosphere above a (non-polarizing) Lambert surface. The observed polarization and geometric albedos cannot be matched simultaneously by such models.

Zellner's three color observations are unique in several respects. The 0.52 μm measurements indicate a steep drop in the polarization from 6° to 4° phase, and the 0.36 μm measurements suggest small negative polarizations near 4° phase. Both characteristics are inconsistent with Rayleigh scattering models.

No cloud model calculations have yet been carried out to match the Titan polarization curves. But Coffeen and Hansen (1973) have analyzed Lyot's white light measurements of the center of the disk of Saturn which resemble the Titan data (cf. Figure 2-18). Thus the conclusions obtained for Saturn can be extended to Titan (Figure 2-21). Cloud models having spherical particles with mean radii of 2-3 μm size distributions $N(r) \propto r^{-2}$ with abrupt cut-offs at 0.75 r_0 and 1.25 r_0 (where r_0 is the mean particle radius), and indices of refraction n = 1.3 to 1.5 at 0.55 μm match the observations (Coffeen and Hansen, 1973).

Summary and Conclusions

The available photometric and polarimetric information about Titan can be summarized as follows:

(1) No changes in brightness or color related to orbital position have been detected. This is consistent with the presence of an optically thick atmosphere;

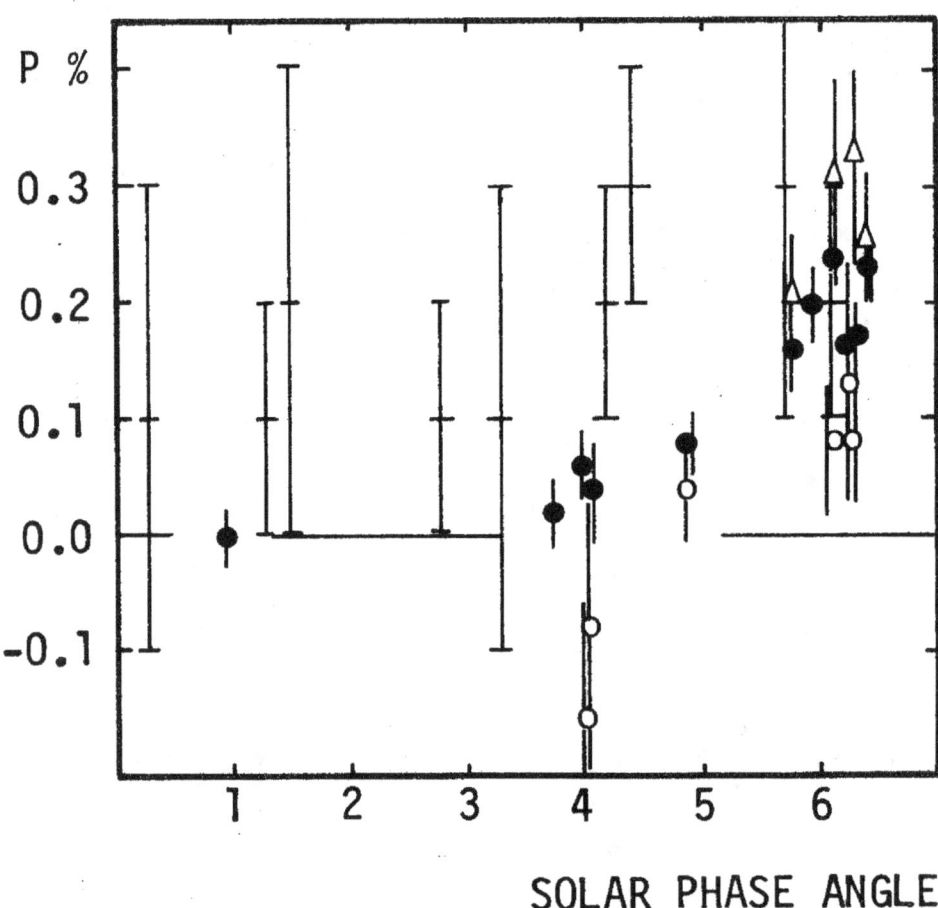

Figure 2-20. Polarization observations of Titan. Open circles indicate observations in ultraviolet light (0.36 μm), filled circles in the
green (0.52 μm), and triangles in the near infrared (0.83 μm),
all made by Zellner in 1970-71. Crosses represent white-light
observations by Veverka (1970) made in 1968-69. After Zellner
(1973). Reprinted from Icarus, 18:662, with permission of
Academic Press, Inc. Copyright © 1973 by Academic Press, Inc.
All rights of reproduction in any form reserved.

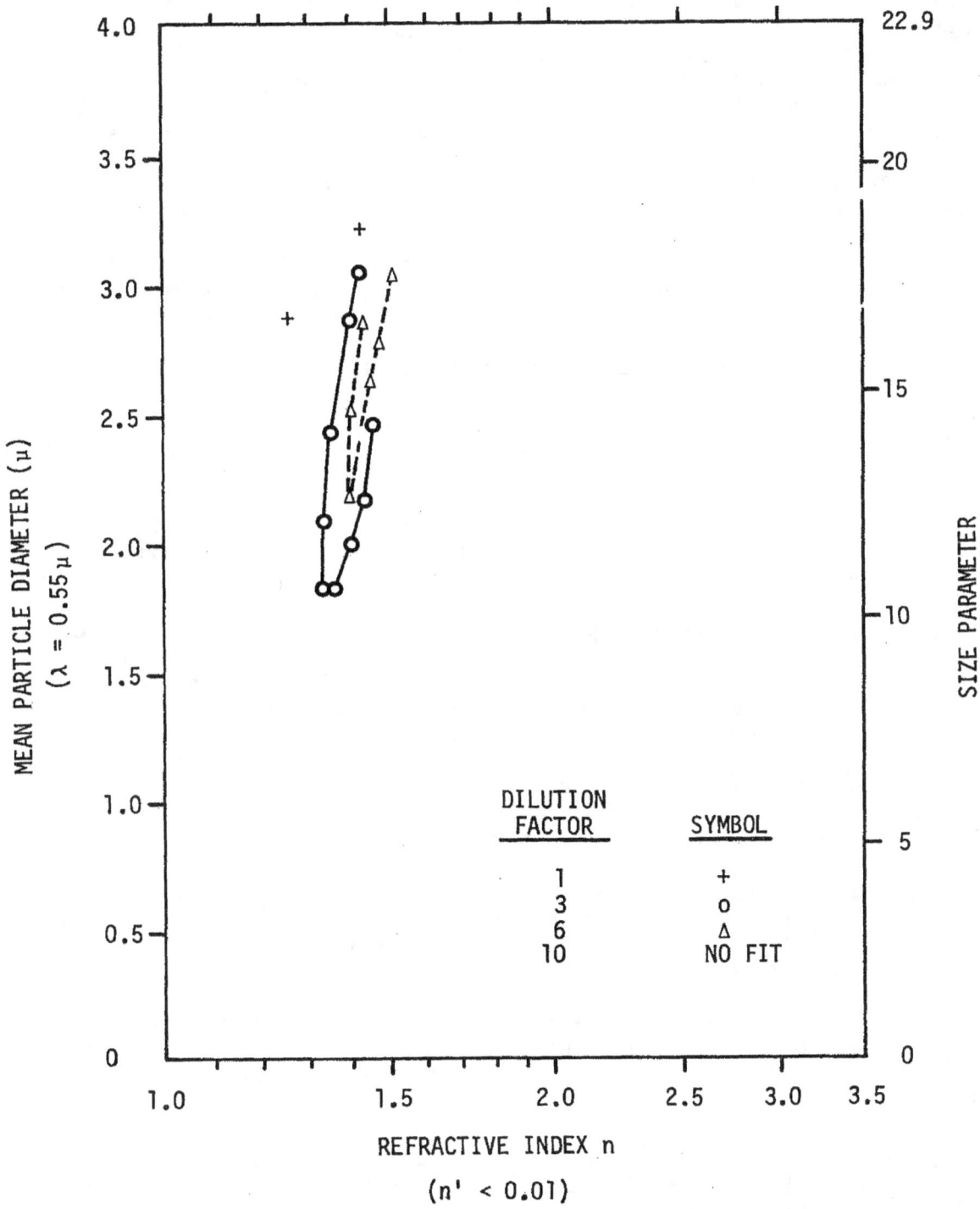

Figure 2-21. Locus of fits of Lyot's observations of the center of the disk of Saturn with Mie calculations for size distributions of spheres, for various multiple scattering dilution factors. After Coffeen and Hansen (1973). Reprinted from Planets, Stars, and Nebulae Studied with Photopolarimetry, copyright © 1974, with permission of University of Arizona Press. All rights reserved.

(2) A phase effect has been detected. The phase coefficients decrease with increasing wavelength from 0.014 ± 0.001 mag/deg at 0.35 μm to 0.001 ± 0.001 at 0.75 μm. The large value of the phase coefficient at 0.35 μm, and its wavelength dependence are inconsistent with an optically thick Rayleigh atmosphere. The low values of the phase coefficients in the red are difficult to explain with optically thin models of the atmosphere. Clouds are required;

(3) The mean opposition magnitude of Titan in the V may have changed from 8.39 in 1961 to 8.35 by 1971, suggesting a possible secular brightening.

(4) The geometric albedo is very low in the UV. Typical values are 0.05 at 0.26 μm and 0.06 at 0.36 μm. This places an upper limit of τ (0.36 μm) $\leq$ 0.04 on the optical depth of any Rayleigh atmosphere above the clouds;

(5) Outside of the methane bands, the geometric albedo may be as high as 0.37, beyond 0.6 μm;

(6) No direct information on the phase integral q exists. Since Titan seems to have an optically thick atmosphere, it is likely that $1 \leq q \leq 1.5$;

(7) Younkin (1973) estimates the bolometric Bond albedo to be 0.27, using $\bar{q}$ = 1.3. Because of the lack of information about q, an uncertainty of ±10% in the Bond albedo is likely (0.27 ± 0.03);

(8) Detailed narrow-band spectrophotometry shows that the spectral reflectance curve of Titan is very similar to that of Saturn's equatorial belt. This suggests a Saturn-like model for the atmosphere of Titan, including an opaque cloud deck. However, the amount of Rayleigh atmosphere above the cloud top must be much smaller on Titan than on Saturn;

(9) The sharp drop in Titan's spectrum below 0.6 μm indicates a strong UV absorber in the clouds. This material may be the same as that responsible for the orange color of the belts of Saturn and Jupiter;

(10) Polarization measurements of Titan are inconsistent with either an optically thick Rayleigh atmosphere, or with an optically thin atmosphere model in which a significant amount of light is scattered from a solid surface. Optically thick clouds are required;

(11) Titan's white-light polarization curve is similar to that of the equatorial region of Saturn. Cloud models with 2-3 μm spherical particles having indices of refraction between 1.3 to 1.5 (at 0.55 μm), are consistent with the observations;

The single most important conclusion to be drawn from the photometry and polarimetry of Titan is that a Saturn-like cloud model may be required to explai the sum of the observations.

Acknowledgement

The author is grateful to C. Sagan and M. Noland for their helpful discussions. This work was supported in part by Grant NGR-33-010-082.

Sagan: Dollfus has reported from visual observations that he sees a changing pattern on Titan which is different from what he sees on other Jovian and Saturnian satellites. How does this tie in with the constant brightness and color data for Titan presented in Table 2-2.

Veverka: I think that the contrast of the changes has been exaggerated. From Lyot's or Dollfus' drawings of Titan, you might expect brightness fluctuations of many percent, whereas, in fact, to about 1% you don't see anything.

Sagan: One other conclusion from the constancy of the brightness of Titan, which I think is important, is that it sets some limit on the existence of breaks in the clouds, if you believe there are clouds. The question of breaks in the clouds is, of course, a critical question for imaging observations of Titan.

Veverka: You probably don't have any very large breaks, but you can have a lot of little ones ....

Sagan: ....which when time averaged always represent the same fraction?

Veverka: Yes.

Danielson: With regard to model interpretations of the phase coefficient variation presented in Figure 2-16, would the observed coefficients be consistent with a snow-covered surface?

Veverka: It is difficult to explain the low values of the phase coefficient beyond 0.6 μm in this way, unless the surface has a normal reflectance of about 0.6 at these wavelengths. Also the snow would have to change its reflectance rapidly with wavelength, being considerably darker at shorter wavelengths. Even then it would be difficult to understand the absence of a negative branch in the polarization measurements at 0.5 μm, since the geometric albedo there is only about 0.21.

Morrison: Younkin's value of 0.27 for the radiometric Bond albedo is substantially larger than the value normally used in the past. Is there some obvious reason why the old values are wrong?

Veverka: One reason is that according to Younkin the geometric albedo of Titan in the near infrared is higher than formerly believed. Also, unrealistic values of the phase integrals have been assumed in the past.

Trafton: We have made independent measurements of the geometric albedo of Titan in the red and our values agree with those of Younkin.

Hunten: You quoted a value of 1.3 for the effective phase integral used by Younkin in his computation of bolometric Bond albedo. Where does this value come from?

Veverka: It is the value that people tend to use for the outer planets and is a reasonable value for a cloud-covered planet. Of course, we don't know what the actual value is.

Morrison: I also have a question about Figure 2-17. If you allow for the fact that the two curves are normalized together and remove all the gaseous absorption in the atmosphere, would you be left with much of an argument that the two curves are similar, except that they're both red?

Veverka: McCord et al. say specifically that what they attach great importance to is the fact that, outside the methane bands, the spectra are similar.

Morrison: Io is red also, yet there is no reason to think the surface of Io is similar to that of Saturn, although perhaps it is.

Veverka: If you remove all the methane bands, all you are saying is Titan is as red as the equatorial belt of Saturn, and so is Io and so are the rings. That's probably a valid argument. The strong absorber of UV light may be the same in all cases. It may occur in the cloud particles on Titan and Saturn, and in the surface layer on Io and the ring particles.

Sagan: Your polarization measurements of Titan in white light show no evidence of a negative branch at small phase angles. Yet Zellner's ultraviolet data do, at least marginally, indicate the presence of a negative branch. Can you explain this difference?

Veverka: That is a perplexing point. If you are saying that the reason you see some negative polarization at 0.36 μm is because you are looking at a surface, you would expect to see more of the surface at 0.8 μm than at 0.36 μm. I think if there is a negative branch at 0.36 μm, it is telling you something about the clouds.

Danielson: What does dust in the atmosphere do? What kind of polarization does that cause?

Veverka: I don't know the answer to that question. It is very hard for me to guess. All I can say definitely is that a pure Rayleigh atmosphere won't do. You need some large scatterers. But what the properties of the scatterers should be, I can only guess. Judging from Coffeen and Hansen's analysis of Lyot's Saturn measurements, which are similar to the Titan data you could explain the Titan observations with a cloud of spherical particles with indices of refraction between 1.3 and 1.5 (at 0.55 μm) and mean particle sizes of the order of two to three microns.

Morrison: Although the spectral reflectance curves of Titan and Saturn are very similar there are differences in the absolute values of the albedos.

Veverka: Yes, and it is important to establish accurately what these differences are. It is hard to measure the geometric albedo of Saturn minus its Rings.

Caldwell: Such differences could be due to differences in the clouds or possibly in the amounts of atmosphere above the clouds. How hard is it to make an ultraviolet polarization observation of Titan? The signal must be very low.

Veverka: Zellner is presently measuring red polarization values to something like plus or minus 0.05%, whereas the ultraviolet values are hard to measure to better than 0.1%.

Postscript, December 4, 1973: Hunten's suggestion (p. 5) that the surface of Titan is being continuously paved by photochemical 'asphalt' falling out of the atmosphere, provides the best means of reconciling the photometric and polarimetric observations with an optically thin Titan atmosphere. This photochemical 'asphalt' would be produced copiously by the action of ultraviolet light on the hydrogen-methane atmosphere in the manner discussed by Strobel, and can probably be identified with the reddish material responsible for the coloration of Titan.

If the production of such substance is efficient and planetwide the surface of Titan may well be covered with a fairly thick, smooth and uniform layer of this material. Then there would be no albedo markings on Titan, which would explain why Titan shows no brightness fluctuations as it revolves about Saturn. Second, this type of surface could be quite smooth optically, making the absence of a negative branch in the polarization curve understandable. Finally, it is likely that such a surface could match the phase coefficients observed for Titan.

Thus Hunten's suggestion makes it possible to have an optically thin atmosphere, but only at the expense of having the surface continually paved by photochemical 'asphalt'. Note that this requires a modification of Danielson's model, because now the reddish aerosol must not only occur in the atmosphere, as postulated by Danielson, but must also copiously cover the surface.

2.5   INTERIOR AND ITS IMPLICATIONS FOR THE ATMOSPHERE

J. S. Lewis

Introduction

Up to now we have been talking about the atmosphere. We all know that the atmosphere represents a negligible proportion of the mass of Titan and that we know very little about it. I'm going to talk about the interior, which represents a much larger proportion of the mass of Titan, and we know almost nothing about it.

Perhaps the best way to proceed is to review conceptually, several contrasting interior models for solar system bodies and then discuss in some detail the thermal history of the model I prefer for Titan. To this I will add some remarks on compatible atmospheric bulk composition and structure. It should prove interesting to see how this blends with the Earth-based atmospheric data presented so far today.

Conceptual Chemical Equilibrium Model

The first thing in dealing with the interior of Titan is to just list those raw materials which might be of some significance in the interior; to draw up a list of those abundant materials which one might expect to be present at that point in the solar system. We need not be either very clever or very well informed to draw up most of the list. We need only to ask what the most abundant elements are and what materials are reasonably easy to condense.

These first results I will be giving are based upon the concept of <u>chemical equilibrium</u> between the solid materials and the gas of solar composition. The simplest way to run through this is to start with high temperatures and say that first rock-forming materials (in other words, silicates, sulfides, and metals) condense. We will assume that by the time we have reached the outer part of the solar system, this entire process of condensing rocks has gone to completion and substantial amounts of less dense and more volatile material have also condensed. This is certainly borne out by the low observed densities of many of the satellites and the low postulated densities of many of the smaller satellites. Morrison might have something to say about the very low densities of some of the satellites of Saturn. Perhaps this is a subject we can come back to later, because I think you'll see that it's a very important point -- whether there are satellites with densities as small as unity.

Below the temperatures at which rocks are formed (again thinking of a sequence of decreasing temperatures, starting with a parent material which has the same composition as the Sun) the next appreciable material to condense is water ice. As the temperature continues to drop, the next major material to form out of this gas of solar composition would be a solid hydrate of ammonia. At yet lower temperatures, methane clathrate hydrate is condensed, which is not strictly a chemically bound compound. This material represents simply sticking a methane molecule into each one of the large vacant sites in the ice lattice. This uses up all the water ice but some methane gas is left over. Further lowering the temperature, the next important thing to condense is the left-over methane, which condenses as a solid.

Now what are the densities of the objects which are formed during this cooling sequence? Densities of rock-like objects, depending upon formation temperature, are on the order of 4 gm cm^{-3}, but there is considerable detail ranging from the condensation of metallic iron (density ~ 7) on down to ~ 3 at lower temperatures. So 4 is just a round number. When water ice is condensed, the bulk density of everything, rock plus ice -- comes out to approximately 1.7 gm cm^{-3}.

I am going to give relative densities which are good to a few hundredths of a gram per cubic centimeter. The absolute densities are not good to anywhere near that accuracy because of the uncertainties in the cosmic abundance of the elements. I shall give three significant figures for the densities, and the ratios of those densities, or the differences between them, are both reasonably secure. However, the absolute densities may be slipped by as much as 0.3 gm cm^{-3} by changes, for example, in the abundance of carbon and oxygen relative to silicon.

Once the ammonia hydrate forms, the mass of condensed material goes up only slightly because the atomic abundance of nitrogen is 5 times less than that of oxygen, and the density would drop slightly. This would give a bulk density of around 1.65 gm cm^{-3}. Next, methane hydrate forms. Because a portion of the water has already been used up in making ammonia hydrate, and because about half of the oxygen has been used up in making silicates, the amount of methane that is retained is not terribly large and the density change is not very large. The bulk density after this step is ~ 1.60 gm cm^{-3}.

Note that these are zero-pressure densities, and certainly compression affects the icy materials quite substantially. For example, if you take the observed density of Titan, 2.2 gm cm^{-3}, and correct it to zero pressure, the density comes out about 1.8. So it's a fairly important correction for objects as large as Titan. Let me also say, parenthetically, that the bulk material which is formed, containing the four constituents down through methane hydrate, contains about 4% methane by weight. Once solid methane is condensed, the amount of methane present goes up by a rather substantial factor to something like 20%.

Notice that carbon is a very abundant material, far more abundant than any of the rock forming elements. It is more abundant than nitrogen, and it has about half the abundance of oxygen. Also, methane has a very low density, about 0.6. That means that, upon methane condensation, the bulk density must drop considerably, and in fact it drops to 1.0 gm cm^{-3}. I hesitate to quote three significant figures here for a simple reason, namely the density is quite temperature-sensitive and quite model-sensitive. Let us say that the zero pressure density is ~ 1.0 after solid methane condenses.

Conceptual Inhomogeneous Accretion Model

Now let us examine a very different but equally tractable formation process. This competing model assumes that the accretion of solid particles into large bodies takes place very rapidly. As soon as the temperatures get low enough for

59

something to condense, it accretes into a large body and is therefore unable to interact chemically with the gas beyond that point. It makes a profound difference in the chemistry because, for example, ammonia hydrate in a chemical equilibrium model is formed by chemical reaction of ammonia gas with ice which already exists, and requires intimate contact between gas and solid. Likewise, methane hydrate is formed by reaction of methane gas and solid ice, and requires contact between gas and solid.

So let's start again with very high temperatures and go through the calculations, but unlike the equilibrium case, as each new condensate appears from the gas, we will remove it. In physical terms, we accrete a layer of the most recently formed material on to the surface of a body which is now inhomogeneously accreted. This is a generalization of the inhomogeneous accretion model of Turekian and Clark (1969) as proposed for the Earth. The core forms first, then the mantle accretes on top of it, then the crust on top of that. Their model has numerous serious difficulties and I am not advocating it; and I don't believe they would for this case either, but it's the farthest removed from the simple, straightforward equilibrium approach that I can conceive of and certainly represents a polar extreme that should be studied.

The sequence begins again with rock, but with certain special differences. No water-bearing silicates are present, because we do not permit water vapor to equilibrate with high-temperature silicates. Also, there are no sulfides, and no iron oxides. This means that sulfur remains in the gas (H_2S) long after the condensation of rocky material is complete.

Then, continuing to cool this gas, which now has a different composition than in the equilibrium example, we would condense water ice first. After that, we would condense a material that does not appear at all in the first sequence. This is ammonium hydrosulfide, which we may also call an ice or a salt. It is a fairly stable solid which has a vapor pressure comparable to ammonium chloride. You can make it and hold it in your hands at room temperature, though it is exceedingly unpleasant. When freshly prepared and pure, it's just as colorless as any ice, or common salt.

Going to yet lower temperatures, ammonia, frustrated from its tendency to react with water to produce solid ammonia hydrate, must condense by itself as solid ammonia ice. This will occur at much lower temperatures than the temperatures at which the ammonia hydrate would normally form. In other words, the vapor pressure of ammonia ice is much higher than the vapor pressure of ammonia hydrate. Finally, going down to even lower temperatures, we get methane ice.

The bulk densities along this sequence look quite similar to those we calculated earlier. Because of the uncertainties in the cosmic abundances, there is no way to use the observed densities of bodies in the outer solar system to distinguish between the two models. Notice, however, that in the inner solar system, there is a very large difference in the chemistry. Just knowing the chemical composition of the Earth or meteorites, knowing the bulk densities of the planets, and knowing a great deal about the structure of the Earth and something (not much) about the internal structures of Mars and Venus, we can compare how the densities of the terrestrial planets ought to vary with distance from the Sun (from the predictions of these two kinds of models) with the observed densities and to the observed compositional detail we have for the Earth. If that is done, the equilibrium model comes out looking very good. It predicts the observed density dependence, including such subleties as the increase in density going outward from Venus to the Earth. That's very rewarding.

I have, however, elected to give you the results of both models for the purpose of equipping you with a complete list of abundant materials that one might worry about as starting materials. By "abundant", I mean those materials abundant enough to effect the bulk density. Those which might be abundant enough to produce a visible trace of atmospheric gas would require pursuing this list down to much, much less abundant elements, because atmospheres represent such a small proportion of the total mass of bodies.

Working Compositional Model for Titan

Now, let us take the equilibrium concept as our working compositional model. We will then assemble these materials into objects comparable in size to the Galilean satellites or to Titan. In other words, large enough so that their internal thermal state is interesting, and they are not simple isothermal objects. Let us sketch how the internal structure of these objects depends upon their composition, and thus upon their formation temperature. The temperatures of formation or condensation for these materials range from about 170°K for water ice to about 120°K for ammonia hydrate, to about 80°K for methane hydrate, and down to about 50°K for solid methane. These condensation temperatures are computed for solar nebula pressures of about 10^{-6} to 10^{-7} atmospheres. Within a factor of 10 or even 100 in pressure, these temperatures simply all shift by a certain logarithmic increment up and down together. The sequence of reactions is immutable. We can, therefore, sketch cross-sections through these condensed objects as a function of the formation temperature, as shown in Figure 2-22.

At 200°K, we have not quite condensed water ice yet, and we may represent the object as being all rock. If we like, we could permit this rock to differentiate by density, to permit the formation of a sulphide core with a silicate mantle on top of it, but this is a matter of taste.

At about 170°K ice will condense, and I will allow the ice to rise to the top via density-dependent differentiations. I will give my reasons for that eventually. With ice added, the structure changes abruptly. The top of the rock region will drop down, and we will have an ice layer on top. The "rock" here means a water-bearing silicate rock, perhaps like serpentine.

As we set the temperature yet lower, ammonia hydrate will be present in the parent material of the satellite. Upon partial melting and differentiation of the satellite the ammonia will partition itself between the various phases and, of course, ammonia is exceedingly soluble in water. The result will be to form a "mantle" of aqueous ammonia solution. A solid mixture of ice and ammonia hydrate begins to melt at a temperature of 173°K which is the eutectic temperature in the ammonia-water system. This means that an object with a surface temperature not much less than 170°K may have a thin ice crust, and, at a relatively shallow depth, the temperatures become high enough so that a melt or slurry (which is a suspension of ice in a liquid) is present.

61

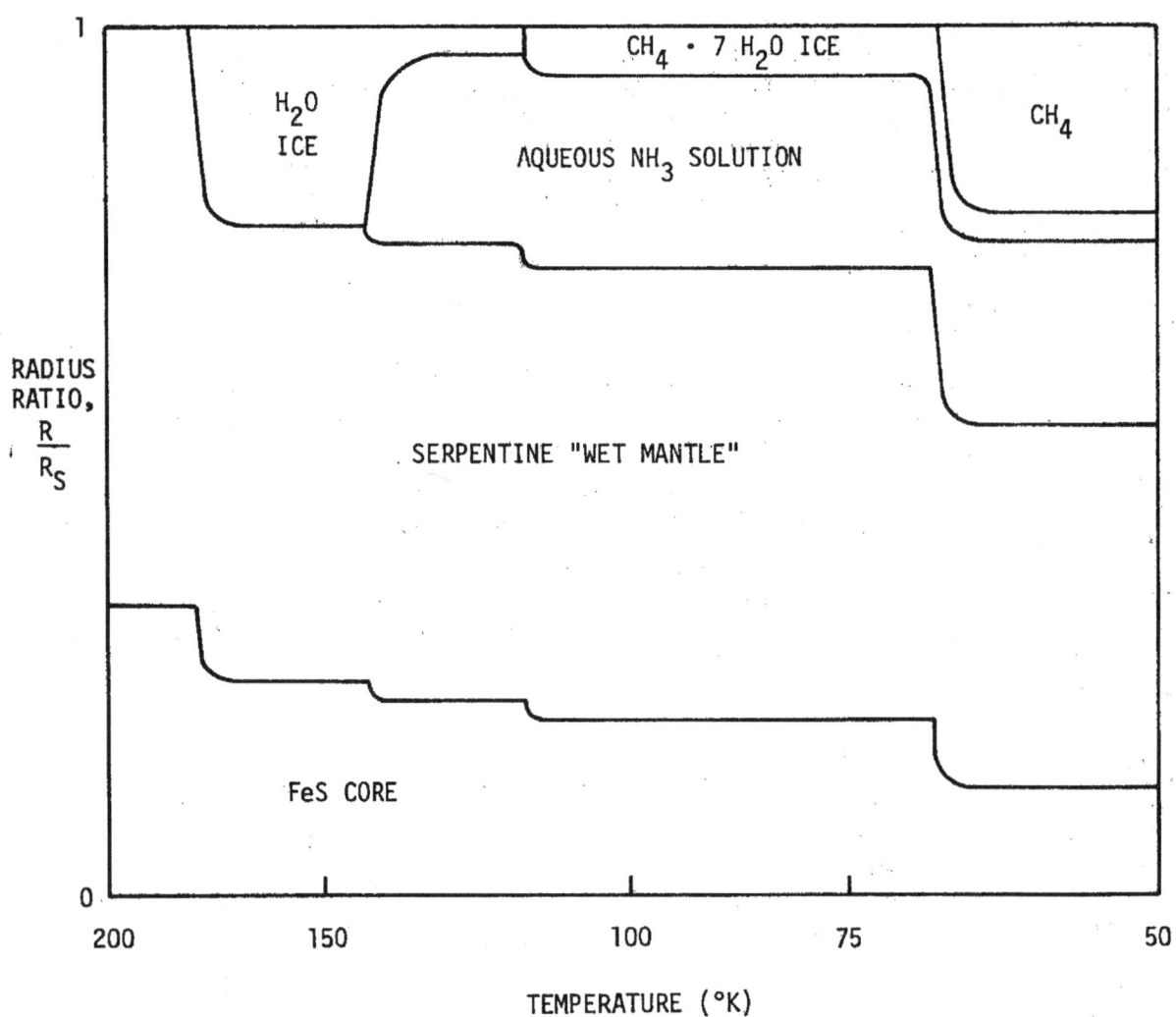

Figure 2-22. Radial sections of condensed objects as a function of formation temperature based on the chemical equilibrium model.

The next step is adding a solid methane hydrate to the parent material. The effect of making solid methane hydrate, you will recall, is to leave left-over methane gas and exhaust all the ice. That means that a differentiated object produced from this composition will contain not only an aqueous ammonia solution, but also a crust which now consists of methane hydrate. Methane gas will be given off due to melting of this ice mixture. Some proportion between zero and 100 percent of the methane will be released as a gas, which will produce an atmosphere.

Then, at temperatures low enough to condense solid methane, a very profound effect occurs, with the superposition of an enormous mass of methane on top of the ice crust. Now, the question is, what is the physical state of this methane? Is it solid, liquid, or methane gas? If a Titan-sized satellite contained such a large quantity of methane, we would find that the physical state of the methane versus depth would depend upon detailed physical considerations, such as the internal thermal structure of the satellite, that we have not yet considered.

An object formed with methane hydrate (but not solid CH_4) present contains enough methane so that, if it were fully outgassed, it would produce a one kilobar methane atmosphere. The total amount of methane obtainable when solid CH_4 is included could reach about 5 kb at the base of the methane, where we have indicated the presence of an ice crust.

It is necessary to point out at this juncture that, until we really know what the "surface temperature" is, we can't be certain that this ice layer will be present. What if the atmospheric mass is so large that it permits a surface temperature above the decomposition temperature of methane hydrate? Then there may be an interface between a methane atmosphere and an aqueous ammonia solution with no crust in between. You can't rule it out. It is an exceedingly interesting possibility.

Right now I am addressing myself to Saturn's system, so I would say the boundary temperature is so low that unless something like the presence of a methane atmosphere raises the surface temperature, you won't get ammonia in the atmosphere. Certainly if the atmospheric mass is large enough, and surface temperatures high enough, to have even aqueous ammonia solution in contact with the atmosphere, then the presence of a significant quantity of ammonia in the lower atmosphere is quite possible.

In the Jovian system, you might indeed have an ammonia-containing satellite, and it might indeed be possible to put ammonia into the atmosphere, but the partial pressure of ammonia would be so low that the atmosphere would be photochemically regulated.

Trafton: What happens to your diagram (Figure 2-22) with all the heats of solidification? Does that upset things?

Lewis: There are two effects. One is that it changes the static structure and the other is that it changes the thermal history.

63

Let me address the first point first. The ice layer we talked about is in fact ice-I only down to a level where the pressure is about 2 kb, at about 50 or 60 km depth. There we get a conversion of ice to a higher density form. Thus there should be a sequence of ice layers going on down to ice-VI. This could be most easily thought of as an isothermal structure in which you just take the isothermal conversion pressures of ice into the high pressure forms.

But now, let's switch over to the second point and ask about a thermal history in which we consider how to get from the homogeneously accreted primitive object, which contains all of these materials mixed together randomly, to a differentiated object. The question then is, what is the time scale for heating? We would like to know, first of all, the formation temperature in order to know what the bulk composition is. We would also like to know the accretion temperature, i.e., the characteristic internal temperature after the solid object is assembled, as a starting point in thermal history models. We would like to know the intensity of heat sources, and, of course, the equations of state and melting behavior of all components in it.

Here is how we do it: We set the accretion temperature equal to the formation temperature. If the object is heated during the accretion process by conversion of gravitational potential energy into internal heat, then it will be easier to melt. So, I am saying, let us make a pessimistic assumption here. Let us set these two temperatures equal. Then, with regard to the intensity of the heat sources, we dismiss gravitational energy as a source of heat. We will do away with adiabatic compression of the interior, and put the thing together isothermally. We will do away with short-lived radionuclides, because there is now some fairly good evidence from the study of meteoroids that the short-lived radionuclides, very popular a few years ago, almost certainly had nothing to do with the thermal history of meteorite parent bodies.

So, we are left with two heating mechanisms to consider, one of which cannot be quantified yet. That is solar wind heating, the Sonett mechanism. The other is long-lived nuclides. But, in the case of long-lived nuclides, we know the abundances of the radioactive elements in meteorites and in the Earth. We, therefore, can make quite reliable estimates of how large that heat source is. The sources of this heat are uranium, thorium and potassium, mostly potassium at this stage of the history of the solar system. The half-life for potassium decay is about a factor of three shorter than that for ^{235}U, ^{238}U, or ^{232}Th, which are about equal-sized heat sources nowadays. Thus, back at the origin of the solar system, potassium was about an order of magnitude more important as a heat source than uranium or thorium.

We are, therefore, taking only one heat source, the one we know must be there, present in the amounts which are needed in order to give the observed density. We want to know how well that heat source, by itself, can heat the parent material, which we start at the lowest possible temperature. For Titan, this is about 70°K. We then ask, how long does it take for that object to heat to the point where it begins to differentiate? Once it begins to differentiate, we have another heat source: the internal conversion of gravitational potential energy to heat, due to this settling of dense material through the light material. This heat source, let us call it ΔGPE, exists only as a result of having started differentiation. Therefore, that heat source has nothing to do with when it starts to melt. Under these assumptions, for an object the size of Titan, it takes $\sim 0.8 \times 10^9$ years to begin to melt.

The additional energy derived from gravitational separation is not by itself sufficient to supply all the heat for phase changes. The Earth differentiated catastrophically because the gravitational potential energy released from separating each part of the Earth was more than enough to melt an equal mass of material. For Titan, this would not be the case. Differentiation would be fairly rapid on a cosmic time scale, but it would not be catastrophic. The differentiation-produced heat source is about four times too small for that to happen. I could probably change that figure by a factor of 2, but, in round numbers, it takes about another 0.8 billion years to do the melting, to reach a steady-state structure. The elapsed time to this point is about 1.6×10^9 years. Thus, within two billion years, Titan would differentiate. Nowadays, it is probably fair to regard the satellites as being in a thermal steady state.

The heat flux that would be passing through the crust of this object to keep it in a thermal steady state is about a thousand times more than the heat flux needed to drive convection in a liquid interior. So heat will be transported rapidly. That means that the liquid mantle will be nearly isothermal. It will be adiabatic in structure, and we will have a thin conductive layer, the crust, on the surface. In the case of Titan, we don't know whether or not the atmospheric mass is large enough to make the surface warm. If it is warm enough, we won't have a crust, and there would be convective heat transport directly up into the base of the atmosphere. I haven't yet said how much atmosphere is needed to get rid of the crust. I said if all the methane were driven out of the interior, it could provide up to a kilobar of methane. It would require <u>at least</u> a 10 b surface pressure to melt and decompose the crust.

Pollack: If there is ammonia in the atmosphere, does your model imply that we are seeing the surface, the aqueous ammonia solution?

Lewis: That depends on what I calculate to be the temperature required to melt the crust. I don't see it necessarily requiring the surface be melted. It may just have ....

Pollack: ....like a volcano?

Lewis: Yes, a volcano, or even solid ammonia hydrate not too far below its eutectic melting temperature.

Pollack: Which one of those two methane compounds would you expect at the surface? If it is a methane hydrate its vapor pressure is a lot lower than that of methane, so if you find a lot of methane in the atmosphere, that implies a certain minimum surface temperature.

Lewis: The vapor pressure of methane over the methane hydrate is several orders of magnitude lower than the vapor pressure of methane over solid methane, so Pollack is saying that the composition of the surface material greatly influences your concept of how the atmosphere interacts with the surface.

Hunten: At 127°K the vapor pressure is 4 bars for methane and 3×10^{-3} bars for the hydrate (Lewis, 1971).

Pollack: You see, this has very strong impact on Danielson's model in the sense that he wants the temperature of the surface to be 80°K and, if that were true, I think it would be very hard to have methane as we observe it in the atmosphere with a clathrate surface.

Danielson: How rapidly do you achieve equilibrium with a clathrate of that kind and if you once sublime it, which means you then form water, what happens? If you take a clathrate and dissociate it, you have water on the surface.

Lewis: Disordered ice, actually, or an aqueous ammonia solution....

Danielson: ....and then you condense out methane again. Then it's just solid methane, isn't it?

Lewis: No, actually, lab experiments were done by Delsemme and Miller (1970) looking at the stability fields of these hydrates to see how relevant they were to comets. They were able to make these hydrates at quite low temperatures I would have to go back to the original article to say exactly how low the temperatures are. They found that granular ice subjected to the presence of methan gas did permit formation of methane hydrate.

Hunten: I have read that this hydrate is an important problem in natural gas pipelines.

Lewis: Yes, it builds up in natural gas pipelines. The gas must be dried quite scrupulously to keep it from happening. It is not just methane; all the light hydrocarbons form these hydrates.

Trafton: For the methane-rich models, what would be the physical character of the surface of the base of the methane?

Lewis: Let me answer that by proceeding to my second and last graph.

Compatible Atmosphere

In this section I will address the relationship of atmospheric composition to bulk composition and interior models. Just for simplicity let us consider a pure methane atmosphere. We'll leave out temporarily, all consideration of ammonia, hydrogen, and other gases. Thinking about a single component is quite a bit easier. Figure 2-23 is the phase diagram for the system, pure methane. Not the triple point of methane at 91°K and 90 mb. By a curious coincidence of natu

66

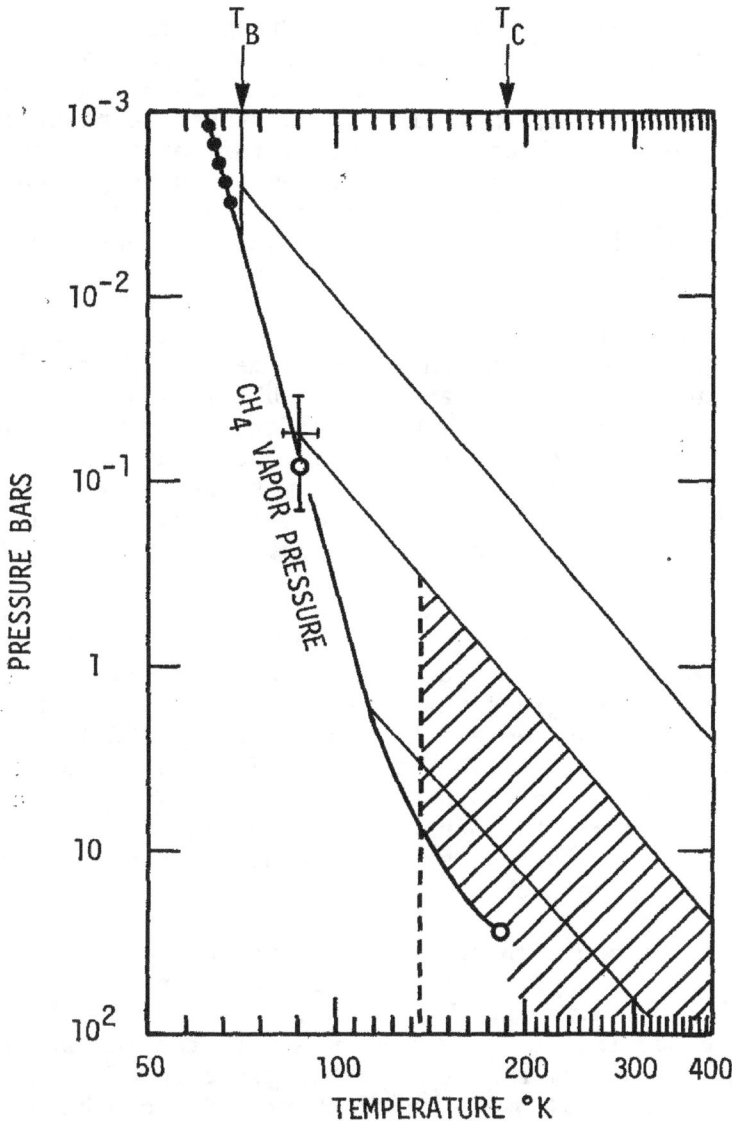

Figure 2-23. Pure-methane atmospheric models for Titan. The heavy line is the vapor pressure curve of methane. T_B is the Gold-Humphreys boundary temperature (74°K), and T_C is the critical temperature of methane. The triple point of methane is indicated at 91°K and 90 mbar, in good accord with the effective temperature and CH_4 pressure for a pure-methane atmospheric model. The three diagonal lines are dry adiabats for pure CH_4. Note that the slope of the vapor pressure curve (a fully-saturated adiabat) equals that of a dry adiabat near the critical point. The cross-hatched region contains the allowable surface conditions ($T_s \geq 145°K$) for pure-CH_4 atmospheres. The visible level in Titan's atmosphere is presumably defined by the triple point of methane, above which level a bright solid-CH_4 particulate haze may be present, but below which only strongly forward-scattering liquid droplets would be stable. After Lewis and Prinn (1973). Reprinted from Comments on Astrophys. Space Phys., 5:4, with permission of Gordon and Breach Science Publishers, Inc. All rights reserved.

the pressures and temperatures observed for Titan correspond fairly well with the triple point of methane. This might be easy to rationalize if we were looking down from above into an atmosphere which contained condensed methane in which the temperature-pressure profile lay, at least temporarily, on the vapor-pressure curve of methane. Thus a thin cloud (or perhaps a thick cloud, depending on one's taste) of solid methane particles would be present down to the level where the melting temperature of methane was reached. Beyond that point, the cloud would be composed of droplets of liquid methane, or the atmosphere could be clear of clouds.

Now what is the real temperature structure of the atmosphere? We do not know, but greenhouse models seem plausible to me, implying an adiabatic lower atmospheric structure. The temperature probably follows the vapor-pressure curve of methane for some distance, but how far? Almost anything could happen. At one extreme, the lower atmosphere may be isothermal; then we are led to predict that there would be a surface of solid methane. Alternatively, the lower atmosphere may have clouds in it down to some greater depth and then the atmosphere may follow a dry adiabat to the surface.

There is a whole family of possible models. If we believe that the temperature of 145°K or 150°K refers to a lower limit on the surface temperature, then we would expect that the surface of Titan would lie to the right of the 150°K isotherm. If one imagines that the lower atmosphere is saturated, then the atmosphere may be saturated all the way to the 150°K level, where the atmospheric pressure is 12 b. On the other hand, 150°K is not an upper limit on the temperature of the surface for this model. It is a lower limit. If the lower atmosphere is saturated all the way down to the critical point of methane, then we may have a methane atmosphere merging gradually into a dense supercritical fluid interior with no phase change.

Alternatively we could follow an unsaturated dry adiabat from our postulated temperature-pressure point in the upper troposphere as far as one cares to do so. Assuming a minimum surface temperature of 150°K then, in the context of the atmospheric models, it means the ones with the lowest atmospheric pressure at the surface of Titan, compatible with the presence of clouds in the upper atmosphere, would have a total pressure of about 0.5 b on the surface.

Regarding models with high surface pressure, I have suggested already that we probably should not think of atmospheric pressures in excess of a kilobar. In the context of these pure methane models, I do not see how one can use present observational evidence to rule out the possibility of having an extremely deep, massive atmosphere. Of course, surface temperatures of about 150°K should inspire those interested in exobiology, because at that temperature ammonia gas appears. Also, 173°K is the eutectic temperature in the ammonia-water system, at which point melting begins.

We therefore must know exactly what that surface temperature is. A radio occultation experiment by Mariner Jupiter/Saturn (MJS) would plumb the atmosphere down to levels where pressures are about 10 b. One should not think that this pressure is significant to two places because the surface temperature is still negotiable. There is quite a variety of features which may be compatible with the notion of an atmosphere with temperature increasing with depth, mostly made of methane which may, by its own presence, make the lowest atmosphere warm enough for ammonia to be present.

Danielson: Why wouldn't your argument apply to some of the Galilean satellites?

Lewis: Well, first of all, they do not have methane atmospheres, which leads us to suspect we are talking about composition classes in which methane was never retained. Secondly, the presence of an equilibrium atmosphere on one of the Galilean satellites would be virtually unobservable. My estimate is about 2×10^{-7} atmospheres, based strictly on chemical equilibrium considerations. Photochemistry will turn over such an atmosphere very quickly. It is noteworthy that the upper limit set by the Io occultation is exactly the same, useless to all concerned. Thus, Io does not have an atmosphere that is capable of affecting the surface temperature.

Trafton: Where would you place Triton?

Lewis: I would place Triton in the solid-CH_4 region. The only thing that convinces me that Saturn's satellites may contain methane is that we see it on Titan. Otherwise, I'd have to conclude that it was a marginal situation. The satellites of Uranus would fall in the same class as Titan. Neptune's satellites should fall in the solid CH_4 region.

Sagan: Triton doesn't have any methane on it.

Lewis: It may, of course, because the boundary temperature is so low that it would all be frozen and thus unobservable. So on Triton, we would be talking about a solid methane surface.

Trafton: What about the density of Triton?

Morrison: John, quote your densities for Triton that you were telling me about.

Lewis: I did a literature search on this. I can't claim it was complete, but I took what I thought to be all of the reliable estimates of the mass and radius of Triton and I came to the conclusion that the density of Triton almost certainly lies between 0.2 and 40 gm cm^{-3}. It is very hard to use this information for subtle compositional discrimination.

Blamont: What about hydrogen and rare gases?

Lewis: I can say a few brief things about noble gases. One is that if we search for a mechanism for retaining helium, the formation of clathrate hydrate wouldn't help because the holes in the ice lattice are so large that helium circulates readily through it. So there is no mechanism for helium retention here.

Sagan: Is that the result of a diffusion calculation?

Lewis: Yes.

Sagan: It's actually calculated?

Lewis: The holes are literally substantially larger than the helium atom. Also, in terms of forming small grains in the nebula, the question arises: what would the helium be doing in the ice lattice in the first place? Because we would form these grains at very low pressures, the amount of helium which might be randomly trapped inside the lattice, just by having a crystal grow around it, would be something like one part in 10^{10}.

Sagan: Occluded helium, I think, can be excluded!

Lewis: Yes, and also absorbed helium. Direct helium condensation requires a temperature which is below 1°K. Since the background temperature of the universe is about 2.6°K, I think we should not take that too seriously. Neon is also too small an atom to form a stable clathrate. Neon should condense eventually as solid neon at a temperature of about 12 to 15°K, which also seems unreasonably low.

Argon forms a clathrate hydrate which is rather less stable than the methane hydrate, and argon is quite a bit less abundant than methane. I did a calculation looking at the equilibrium partitioning of argon between gaseous methane and solid methane hydrate and came to the conclusion that one argon atom would be present in the solid for every 7,000 methane molecules. There are certain systematic uncertainties in deriving that number, but the number is not temperature sensitive because the vapor-pressure curves of the hydrates are quite nicely parallel.

Rasool: How about Argon-40?

Lewis: Radiogenic Argon-40, produced in the deep interior, might possibly make it to the surface particularly if we believe that potassium resides in solution rather than in mineral grains. The maximum amount of Argon-40 would be about 10 millibars.

Blamont: That may not be insignificant.

Lewis: If the total pressure is some tens of millibars then it might be quite significant. Finally, getting back to your question about other gases, hydrogen condenses at about only 7°K. Adsorption of gases on solid surfaces is negligible at temperatures more than about 3 times the condensation temperature of the pure substance, so, if we accept $\sim$70°K as the accretion temperature of Titan, no ad-sorbed hydrogen will be retained.

Veverka: Can we get back to Saturn's other satellites? Did you say the other satellites should have solid methane surfaces?

Lewis: I said that, if they had a density of 1, then they would have enormous quantities of methane sitting on top of them. From outside, we would see gaseous methane, I should suspect, and they should have solid surfaces. Frankly, I am skeptical about such low densities.

Veverka: What actual surface material do you suspect?

Lewis: Asphalt.

Morrison: The inner satellites cannot be covered with a material as dark as asphalt unless either the measured masses are off by an order of magnitude or the densities are a great deal less than unity.

Conclusions

The bulk composition and interior structure of Titan required to explain the presence of a substantial methane atmosphere are shown to imply the presence of solid $CH_4 \cdot 7H_2O$ in Titan's primitive material. Consideration of the possible composition and structure of the present atmosphere shows plausible grounds for considering models with total atmospheric pressures ranging from $\sim$20 mb up to $\sim$1 kb. Our expectations regarding the physical state of the surface and its chemical composition are strongly conditioned by the mass of atmosphere we believe to be present. A surface of solid CH_4, liquid CH_4, solid CH_4 hydrate, H_2O ice, aqueous NH_3 solution, or even a non-surface of supercritical H_2O-NH_3-CH_4 fluid could be rationalized. It is an urgent necessity to determine the location of the surface of Titan; in other words, to find the surface atmospheric pressure.

Note: This article is, in part, a summary of publications and preprints by Lewis (1971, 1973), and Lewis and Prinn (1973).

2.6   MODELS OF TEMPERATURE STRUCTURE AND GENERAL CIRCULATION

J. B. Pollack

Rival Models of Titan's Atmosphere

Most of this paper will deal with greenhouse models of Titan's atmosphere. Towards the end, I will summarize some work done by Leovy and myself, in which elementary, general circulation models of its atmosphere are considered. This latter subject has significance both for providing estimates of horizontal temperature variations and in illuminating another unique characteristic of Titan's atmosphere.

Up until recently, infrared brightness-temperature measurements of Titan have been interpreted within the context of greenhouse models. Such models imagine that the atmospheric pressure is sufficiently high so that the lower atmosphere is opaque over almost all wavelengths containing large amounts of thermal radiation. Accordingly, the surface temperature is much higher than the effective temperature at which the satellite radiates to space and the lower atmosphere is characterized by a decrease of temperature with an increase in altitude. High infrared brightness temperatures are interpreted as occurring at wavelengths of reduced atmospheric opacity, which permit penetration to warmer temperature levels.

An alternative model has been proposed by Danielson, et al. (1973). According to this view, the upper atmosphere is heated to a high temperature through the absorption of much of the solar energy reaching the satellite and the surface has a temperature close to the effective temperature. In this case, high infrared brightness temperatures are understood as occurring at wavelengths that are fairly opaque, permitting a view of the warm upper atmosphere. Danielson's model considers almost all of the atmosphere to be at a high temperature, with only a narrow region next to the surface serving as a transition region between the two temperature extremes.

Danielson's model is partially supported by recent narrow-band measurements made in the 8 to 13 micron region by Gillett et al. (1973). As detailed below, at least part of these observations do refer to a warm upper atmosphere. It therefore seems appropriate to reconsider my greenhouse models, (Pollack, 1973), which were based on broad-band infrared measurements.

There are two questions that need to be addressed and these are quite separate from one another. The first concerns itself with the origin of the high brightness temperatures found in the 8 to 13 micron region. There is a continuum of possible answers to the first question: At some wavelengths the radiation may arise from a hot upper atmosphere and, at the remaining wavelengths, from a hot lower atmosphere. The ends of this continuum are the pure Danielson model and the pure greenhouse model.

The second question concerns itself with whether or not the surface has a temperature significantly above the effective temperature. In principle, one could imagine a situation in which all the 8 to 13 micron radiation came from a hot upper atmosphere and yet there was a significant greenhouse effect in the lower atmosphere. For example, clouds could prevent our viewing the lower atmosphere at wavelengths shortward of 13 μm. The answer to this second question may be crucial in assessing the feasibility of a Titan Entry Probe mission. If there is a significant greenhouse effect, the surface pressure would be high enough, so that measurements from an Entry Probe could commence well above the surface. This situation would not necessarily hold if Danielson's model is correct.

Greenhouse Model Revisited

Let us now review my prior greenhouse calculations. Because of Titan's low effective temperature, the only way to achieve a large greenhouse effect is by means of pressure-induced absorptions. Figure 2-24 illustrates the absorption coefficient of the pressure-induced transitions of hydrogen and methane as a function of wavenumber. The notation X → Y indicates that molecule X is absorbing in the presence of molecule Y. To cause a significant greenhouse effect the wavenumber region from about 50 cm^{-1} to 700 cm^{-1} should be opaque. We see from Figure 2-24 that in an atmosphere containing comparable amounts of methane and hydrogen, as is perhaps suggested by the spectroscopic observation, the methane and hydrogen opacity complement one another quite nicely. Where the hydrogen absorption is weak, the methane absorption is strong and vice versa.

Figure 2-25 shows a determination of the pressure-temperature structure of an atmosphere containing equal amounts of hydrogen and methane and having a surface temperature of 150°K. We see that the atmospheric temperature begins to rapidly increase with pressure above a pressure level of 0.1 atm. Put another way, in an atmosphere of this composition, if the surface pressure is above 0.1 atm, there will be a significant greenhouse effect. The spectroscopically determined gas amounts of methane and hydrogen indicate pressures of several hundredths of an atmosphere. The amounts of methane and hydrogen are comparable, as long as there is no other major component to the atmosphere, such as nitrogen. Trafton has shown that much of the line formation takes place in the vicinity of a cloud layer. Therefore, the surface pressure is significantly greater than several hundredths of an atmosphere and is close to or above the critical value of 0.1 atm needed for a greenhouse effect to become important.

Using model atmospheres of the type shown in Figure 2-25, I calculated the dependence of brightness temperature on wavelength. Figure 2-26 illustrates this dependence for atmospheres containing only methane and hydrogen. The curves are labeled according to the value of the hydrogen mixing ratio assumed. Also shown as circles are the four broad-band observations, available at the time the calculation was performed. The horizontal bar through the circle represents the band pass of the instrument used, while the vertical bar indicates the estimated error in the brightness temperature value.

The 20-micron observation provides a good discriminant of the hydrogen mixing ratio for this set of model atmospheres. This point is further illustrated in Figure 2-27, where the theoretical spectra have been convoluted over the instrument band pass at 20 μm. We see that approximately even proportions of hydrogen and methane are implied by the observed brightness temperature.

Table 2-5 summarizes the broad-band observations that have currently been made, as well as the recent narrow-band results of Gillett, et al. (1973). The broad-band values used for Figure 2-26 are marked with a star. The narrow-band observations force us to reconsider our interpretation of the 8-13 micron spectra.

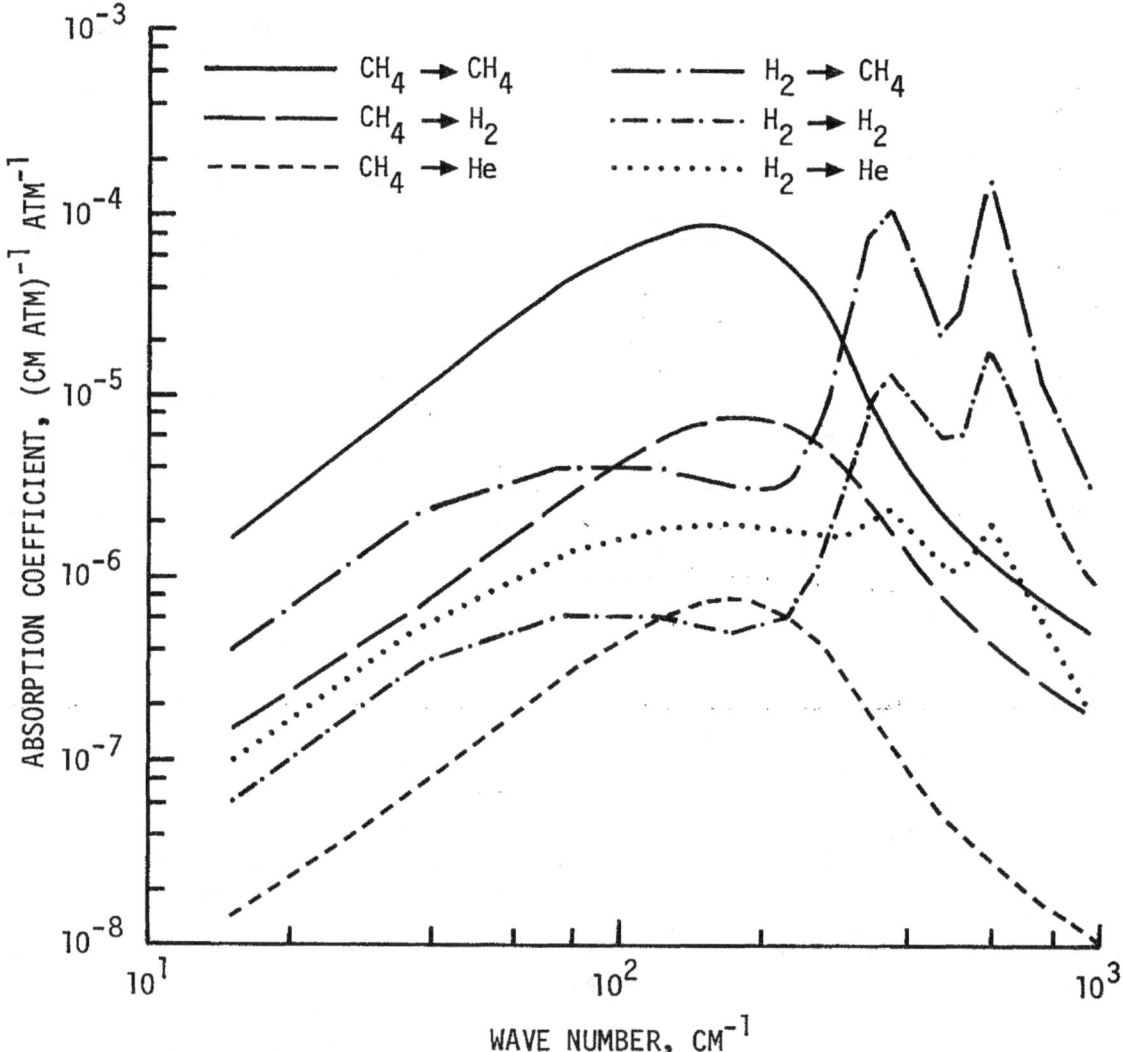

Figure 2-24. The absorption coefficients of the pressure-induced transitions of methane and hydrogen as a function of wavenumber. The notation X → Y indicates that molecule X is absorbing in the presence of molecule Y. After Pollack (1973). Reprinted from _Icarus_, 19:48, with permission of Academic Press, Inc. Copyright © 1973 by Academic Press, Inc. All rights of reproduction in any form reserved.

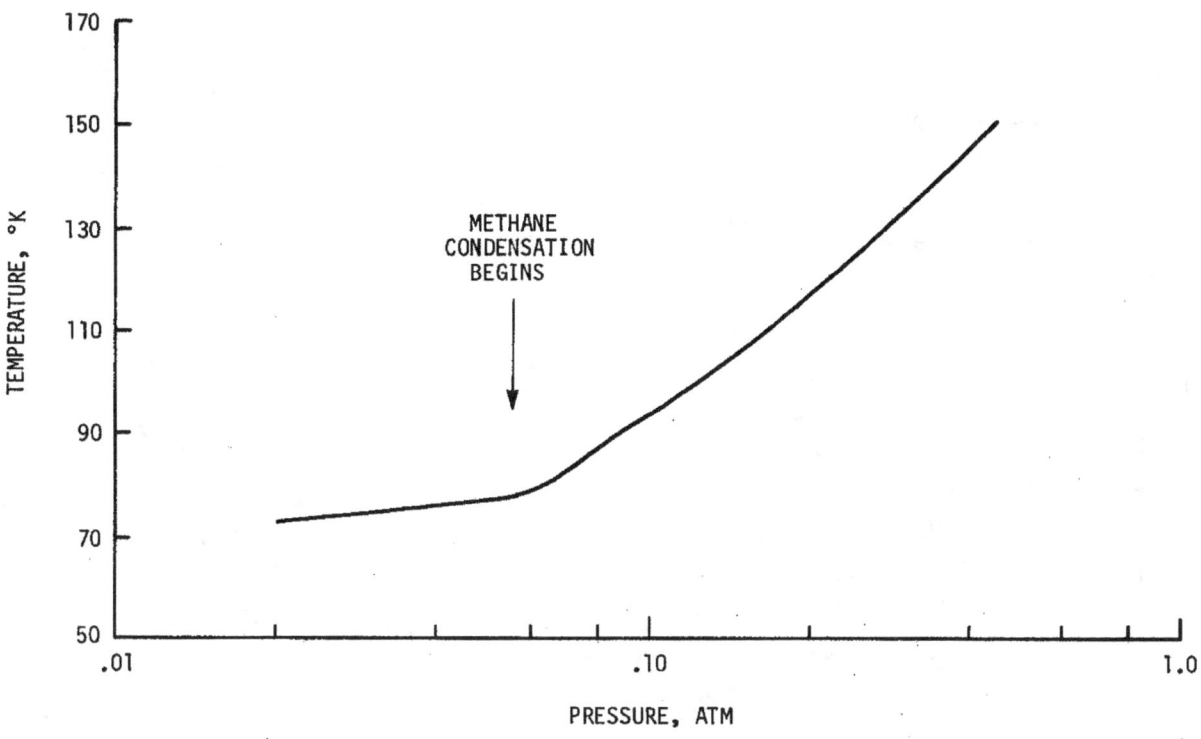

Figure 2-25. Temperature as a function of pressure for a model atmosphere of Titan containing equal amounts of hydrogen and methane and no helium. The mean surface temperature is 150°K. After Pollack (1973). Reprinted from Icarus, 19:49, with permission of Academic Press, Inc. Copyright © 1973 by Academic Press, Inc. All rights of reproduction in any form reserved.

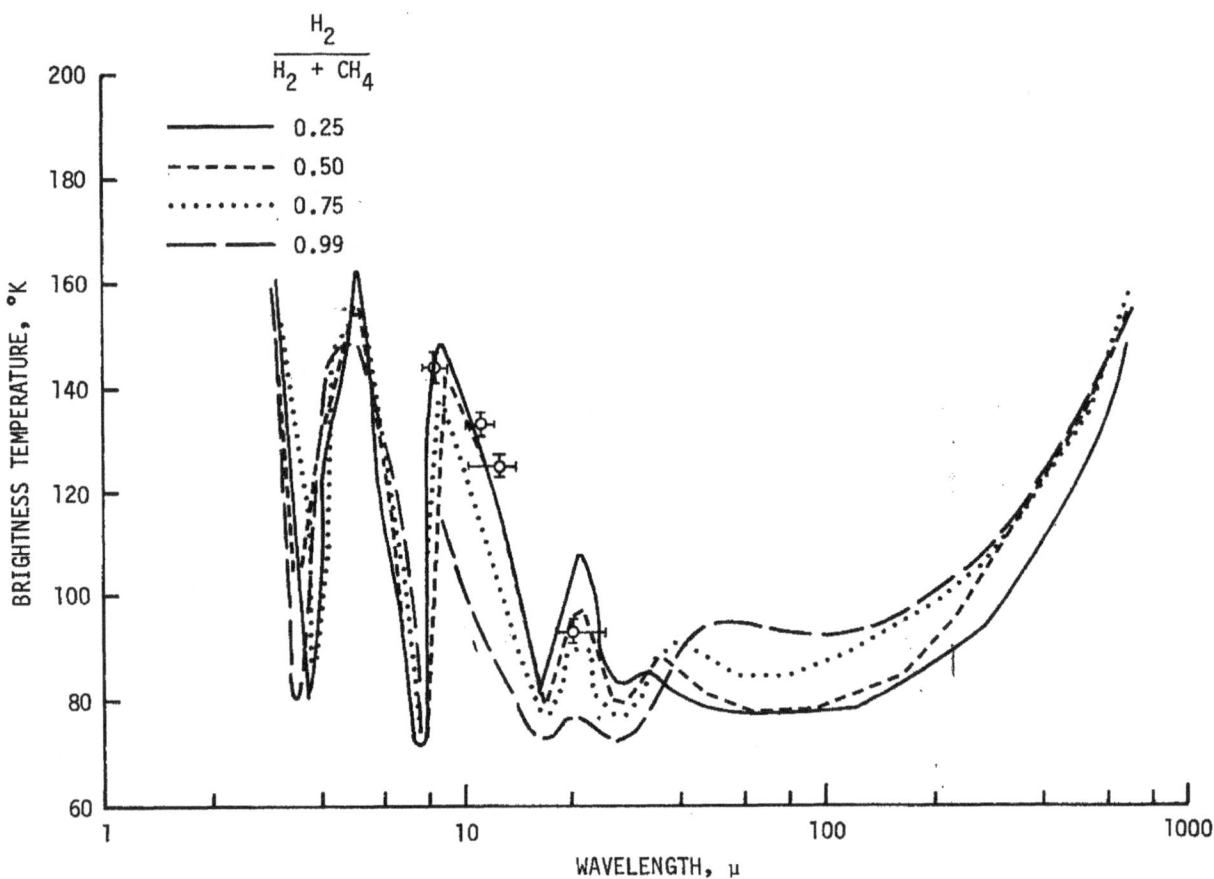

Figure 2-26. Brightness temperature as a function of wavelength for model atmospheres containing no helium and ammonia and varying proportions of methane and hydrogen. Observed temperatures are indicated by a circle, with vertical bars indicating the bandpass of the instrument. After Pollack (1973). Reprinted from Icarus, 19:50, with permission of Academic Press, Inc. Copyright © 1973 by Academic Press, Inc. All rights of reproduction in any form reserved.

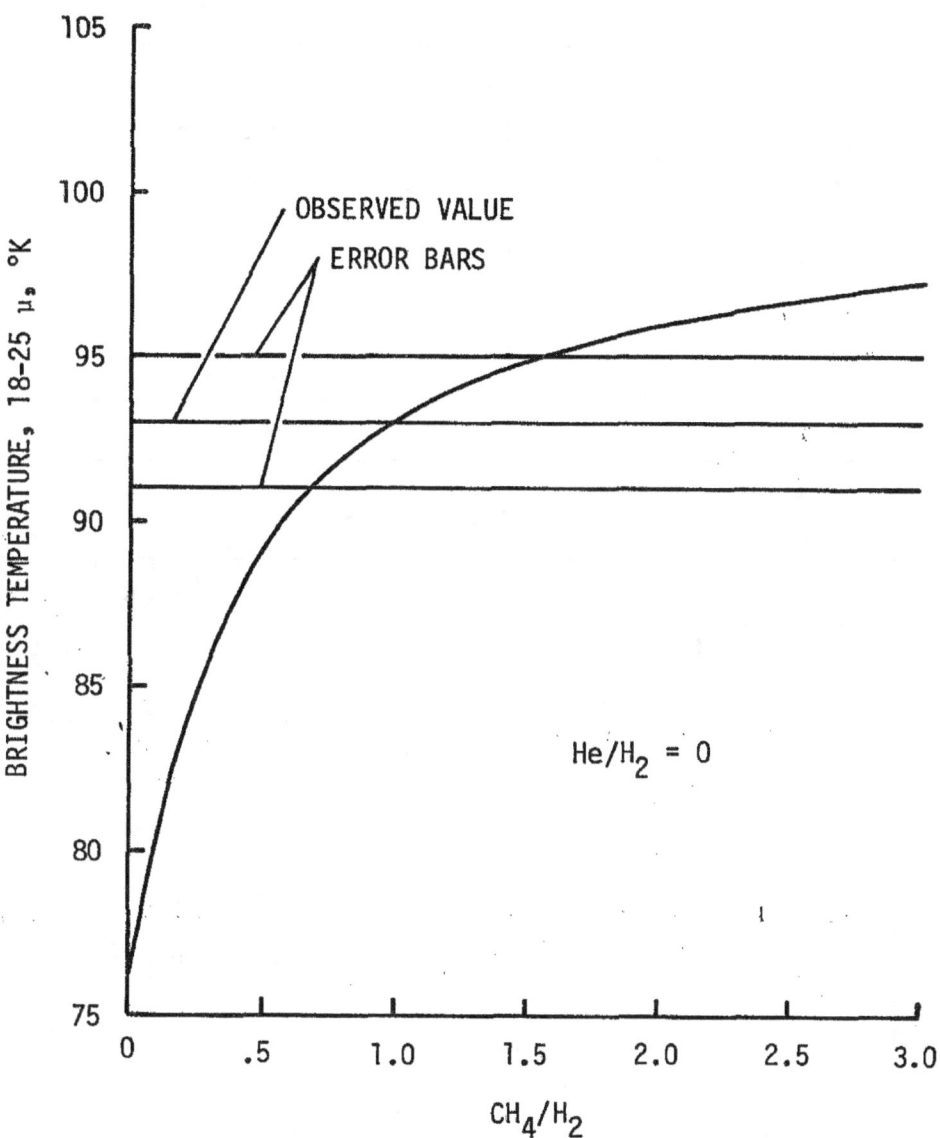

Figure 2-27. Theoretical brightness temperatures applicable to the 20 μm observation as a function of the methane-to-hydrogen ratio. The observed value with its mean and extreme values, permitted by the quoted error bars, is indicated by the horizontal lines. No helium is present in the model atmospheres. After Pollack (1973). Reprinted from Icarus, 19:50, with permission of Academic Press, Inc. Copyright © 1973 by Academic Press, Inc. All rights of reproduction in any form reserved.

Table 2-5. Observed Brightness Temperatures of Titan

Narrow-Band Measurements (Gillett, _et al_. 1973)

λ	T_b (°K)
8.0	158 ± 4
9.0	130 ± 6
10.0	124 ± 3
11.0	123 ± 3
12.0	139 ± 1
12.5	129 ± 2
13.0	128 ± 2

Broad-Band Measurements

λ	Δλ	T_b (°K)	AUTHORS
4.9	0.8	<190	Joyce, _et al_. (1973)
8.4	0.8	146 ± 5	*Gillett, _et al_. (1973)
10.0	5.0	132 ± 5	Low (1965)
11.0	2.0	134 ± 2	*Gillett, _et al_. (1973)
12.0	2.0	132 ± 1	Gillett, _et al_. (1973)
12.4	4.0	125 ± 2	*Allen & Murdock (1971)
20.0	7.0	93 ± 2	*Morrison, _et al_. (1972)

* Used in prior calculations.

The measurement at 8.0 μm lies well within a very strong methane band centered at 7.7 μm. For the type of atmosphere discussed above, containing approximately equal proportions of hydrogen and methane, optical depth unity at 8.0 μm is reached at about the 10^{-3} atm level. This inference was based on laboratory measurements of the 7.7 μm methane fundamental. Thus we are led to conclude that the upper atmosphere is quite warm and that at least some of the radiation in the 8 to 13 micron region comes from the upper atmosphere. In addition, the apparent local peak in brightness temperature near 12 μm may be due to radiation within a strong band of ethane, as suggested by Danielson, et al. (1973).

According to Danielson's model, the remainder of the high brightness temperature radiation in the 8 to 13 micron region should be attributed to a warm upper atmosphere, with the opacity being supplied by a diffuse haze layer located throughout the atmosphere, which has been photochemically generated. The haze is not optically thick at these wavelengths and its opacity is declining with increasing wavelength. Thus, in this model there is some contribution from the cool surface.

However, the remaining portions of the 8 to 13 micron region can equally well be understood within the context of a greenhouse model. According to Figure 2-26, the brightness temperature near 12.5 μm should be about 115°K; this prediction is consistent with the observations if we allow for the rather sizeable uncertainties in the shape of the far wings of the hydrogen rotational lines, which are responsible for the 12.5-micron opacity. These wings have been measured only for $H_2 \rightarrow H_2$, not for $H_2 \rightarrow CH_4$. The latter is by far the more important for Titan. Gillett's 13-micron observation is inconsistent with Danielson's model. However, it may not be inconsistent with some variant of this model.

Figure 2-28 shows the brightness temperature spectra for atmospheres containing no ammonia in the lower atmosphere and amounts of ammonia dictated by its vapor pressure curve. The latter situation corresponds to the maximum amount of ammonia that could be present. We see that the observed narrow-band brightness temperatures, as given in Table 2-5, are close to the values expected for the saturated ammonia model. Thus these data points may indicate the presence of small amounts of ammonia in the lower atmosphere.

The dependence of brightness temperature on wavelength for a variety of surface temperatures is illustrated in Figure 2-29. These calculations once again refer to greenhouse models containing equal proportions of hydrogen and methane. They illustrate that there is some residual opacity at 12.5 μm and hence surface temperatures in excess of the observed brightness temperature of 129°K would be required, within the context of this model.

Finally, in Figure 2-30, we show how the surface pressure varies when we introduce helium into the model atmosphere. The surface temperature has been fixed to a value of 155°K and the atmosphere contains equal proportions of hydrogen and methane. As the helium content is increased, the surface pressure increases. For the case of zero helium, perhaps the most likely a priori case, a surface pressure of 0.45 atm is required.

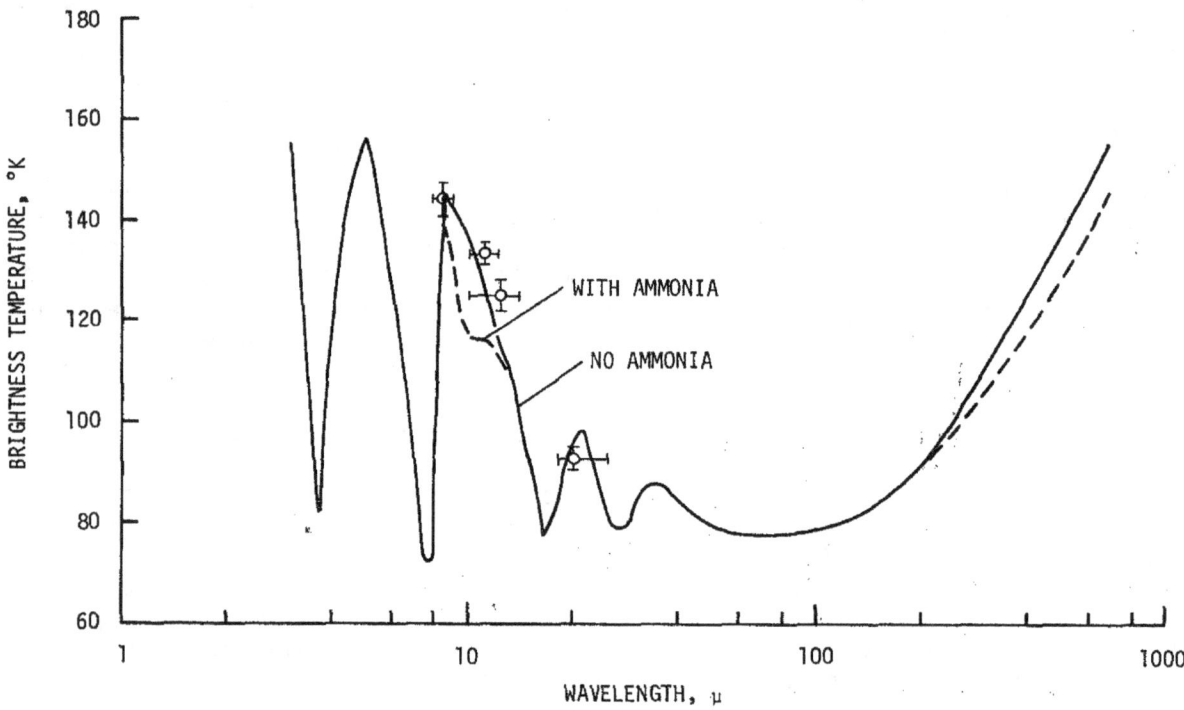

Figure 2-28. Brightness temperature as a function of wavelength for model
atmospheres with and without ammonia. Both model atmospheres
have equal proportions of hydrogen and methane, no helium, and
a daytime surface temperature of 155°K. Observed temperatures
are indicated by a circle, with vertical bars indicating the
estimated error and horizontal bars the band-pass of the
instrument. After Pollack (1973). Reprinted from Icarus,
19:51, with permission of Academic Press, Inc. Copyright
© 1973 by Academic Press, Inc. All rights of reproduction in
any form reserved.

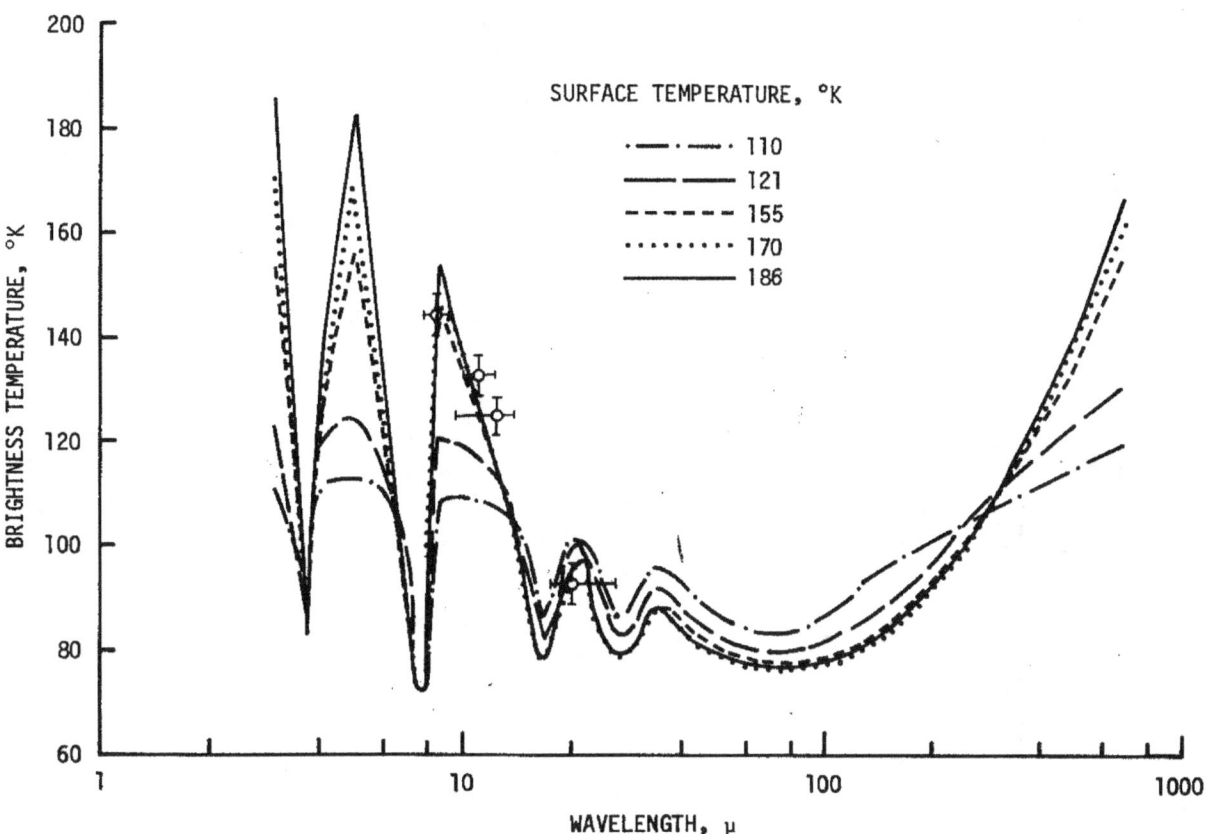

Figure 2-29. Brightness temperature as a function of wavelength for model atmospheres having no helium or ammonia and equal proportions of methane and hydrogen. The surface temperature is varied between models. The observed temperatures are indicated by a circle, with vertical bars indicating the estimated error and horizontal bars the band-pass of the instrument. After Pollack (1973). Reprinted from Icarus, 19:53, with permission of Academic Press, Inc. Copyright © 1973 by Academic Press, Inc. All rights of reproduction in any form reserved.

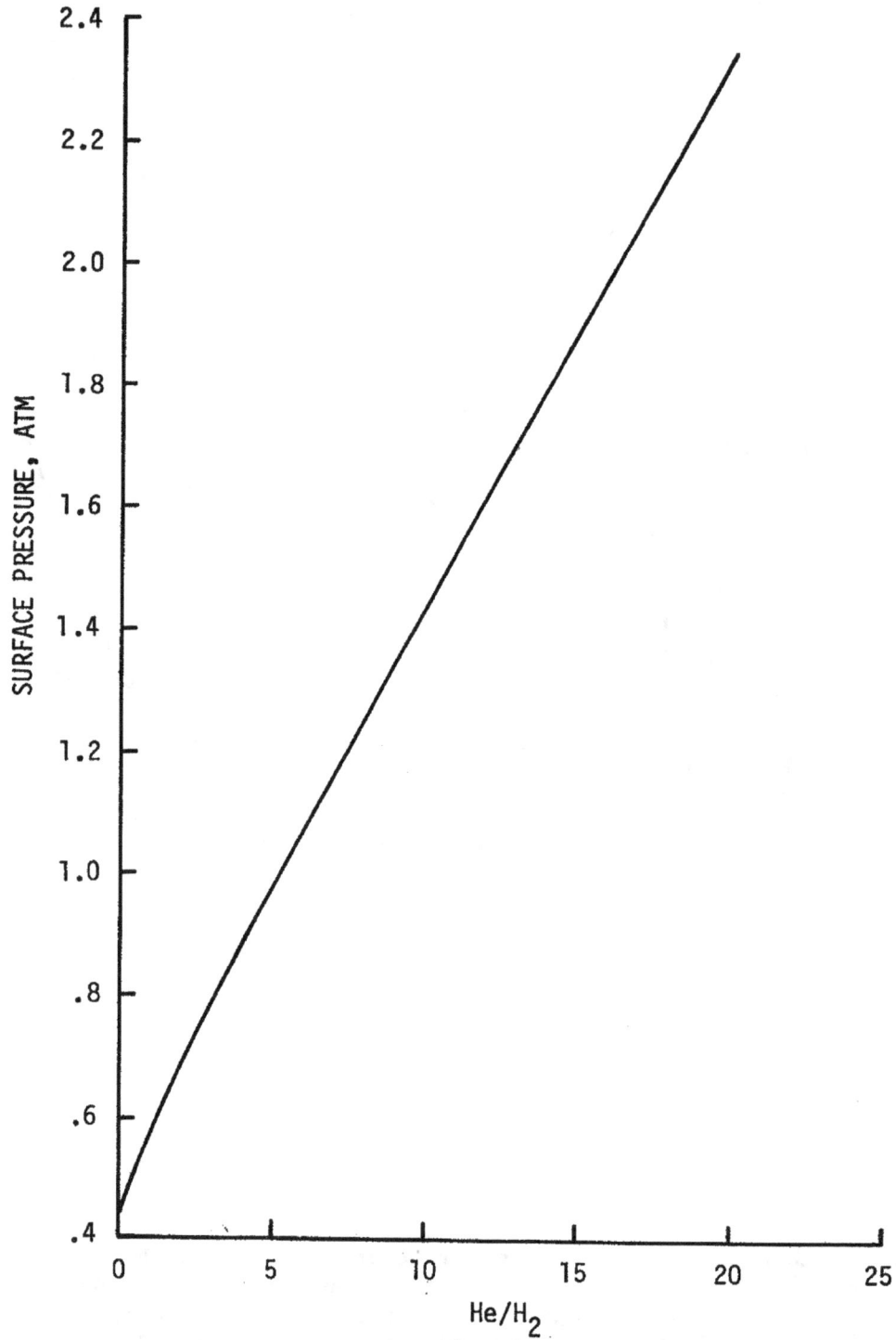

Figure 2-30. Surface pressure as a function of the helium-to-hydrogen ratio for model atmospheres having equal proportions of methane and hydrogen and a surface temperature of 155°K. After Pollack (1973). Reprinted from Icarus, 19:53, with permission of Academic Press, Inc. Copyright © 1973 by Academic Press, Inc. All rights of reproduction in any form reserved.

Future Observations

Above we have seen that it is not possible to deduce a unique model of the atmosphere of Titan from the presently existing data. At some of the wavelengths in the 8 to 13 micron region, e.g., 8.0 μm, the radiation undoubtedly originates from a warm upper atmosphere; however at other wavelengths, e.g., 10, 11, 12.5, and 13 μm, the radiation may originate either from a warm upper atmosphere or a warm lower atmosphere. The distinction between lower and upper atmosphere as used here is that the lower atmosphere is envisioned as optically thick at all important thermal wavelengths, i.e., 15 μm < λ < 200 μm, while the reverse is true of the upper atmosphere. Even should future observations in the 8 to 13 micron region show that all the radiation in this region comes from a hot upper atmosphere, the possibility of a hot lower atmosphere could not be dismissed.

Future spectral observations of Titan at wavelengths longward of 15 μm can determine whether the lower atmosphere and surface have significantly higher temperatures than the effective temperatures. Radio observations permit a direct determination of the surface temperature and are discussed elsewhere in this report. Here we consider the significance of future spectral infrared observations beyond 15 μm. As can be seen from Figures 2-24 and 2-26, greenhouse models predict the presence of several spectral features in this wavelength region. In particular, there should be comparatively sharp features centered at 17 and 28 μm due to hydrogen pressure induced transitions and a broader feature centered close to 50 μm due to the sum of the methane pressure-induced transitions. Whether these features show up as absorption or emission features depends on the exact temperature structure of the atmosphere. In this regard, Figure 2-26 should not be taken literally. However, if the greenhouse model is correct, these features should be present. Groundbased measurements within the 20-micron region will permit a search for part of the structure expected from the hydrogen transitions, while observations from the C-141 aircraft will permit a view of the entire spectral region of interest.

Clouds

According to Danielson's model, aerosols produced photochemically should be present throughout the entire atmosphere. In addition, several types of condensation clouds may also be present. The model atmosphere shown in Figure 2-25 becomes cold enough in the upper part of the troposphere for methane to condense out. This occurs near the 5×10^{-3} atm pressure level. However, whether such clouds will indeed exist in the atmosphere of Titan depends on the details of the solar energy deposition profile. The model atmosphere calculated in Figure 2-25 was based upon the assumption that all the solar energy is deposited at the surface. The existence of a warm upper atmosphere, as discussed above, may mean that the tropopause is too warm to permit methane condensation anywhere in the atmosphere. However, it is worth noting that several independent pieces of evidence, discussed elsewhere in this report, suggest the presence of clouds with a definite bottom, located near the pressure levels expected for the methane clouds.

We have also seen in this discussion that the new 10 and 11 micron narrowband temperatures may imply the presence of ammonia in the lower atmosphere. At the temperatures appropriate to Titan ammonia is severely limited by its saturation vapor pressure curve and, hence, ammonia clouds may be present in the lower atmosphere. The possible presence of ammonia in the lower atmosphere needs to be factored into considerations of potential constraints on the communication link between a Titan Probe and its Relay Bus.

Solar Energy Deposition

The above greenhouse calculations have been performed assuming that all the solar energy deposition takes place at the surface. It is quite legitimate to inquire whether these results will significantly be modified if one allows for large amounts of solar energy deposition within the atmosphere, because of absorption by both the near infrared vibration bands of methane and those of the cloud aerosols. A somewhat analogous situation exists for Venus, where a bright cloud layer in the upper troposphere absorbs some of the incident sunlight and the atmosphere absorbs a significant fraction of the sunlight penetrating beneath the clouds. Yet the recent Venera-8 Probe has shown that some sunlight does reach the surface and is probably responsible for the high surface temperature in the sense of a classical greenhouse effect. This situation is also consistent with greenhouse calculations I have performed for Venus, which indicate that even large amounts of solar energy absorption within the atmosphere do not significantly change the greenhouse effect. We conclude that if Titan has a massive enough atmosphere (say, a surface pressure greater than 0.1 atm), it will have a significant greenhouse effect.

Atmospheric Dynamics

Scaling the equations of motion, Leovy and Pollack (1973) have made a first estimate of the atmospheric dynamics of Titan. They considered both baroclinic and axially symmetric general circulation models and concluded that the latter was much more probable. Titan's circulation may represent an important intermediate case between the baroclinic circulation typical of mid-latitude regions on the Earth and Mars on the one hand and the symmetric circulation of Venus on the other hand. Titan is rotating slowly enough to have a symmetric circulation. Yet in contrast to Venus, which rotates much slower, coriolis effects are probably quite important for Titan. Conceivably by studying Titan's circulation, as might be possible with a multiprobe, second generation mission to Titan, we might obtain important insights into the circulation of the Earth's equatorial regions.

A second important aspect of Leovy and Pollack's study is their estimates of horizontal temperature gradients within Titan's atmosphere and along its surface. Let us first consider the atmosphere. The diurnal variation is severely restricted by simple considerations of thermal inertia. Assuming that the atmosphere is optically thick at all thermal wavelengths, that the surface pressure is 1/3 of a bar, and that the period of rotation equals the orbital period, they find a diurnal temperature amplitude of 0.03°K. To first order this result will scale inversely as the surface pressure and so will remain quite small, even for the smallest surface pressure considered likely ($\sim 2 \times 10^{-2}$ atm). Equator to pole temperature variations are limited by the atmospheric meridional circulation. For the same conditions given above, Leovy and Pollack estimate a latitudinal temperature variation on the order of 0.1°K. This variation scales approximately as the surface pressure to the (-5/3) power. As a result, this variation could be important at the lowest possible values of surface pressure (~ 10 mb).

We next turn to Leovy and Pollack's estimates of surface temperature variations. In the case of a strong greenhouse effect, the atmospheric radiation will severely restrict these temperature variations. For example, with a mean surface temperature of 150°K, the surface temperature amplitude is less than 15°K. In the case of Danielson's model, radiation from the atmosphere also serves to prevent extremely large temperature variations. However, the atmosphere, which in this model radiates effectively only on the Wien tail of the blackbody function, may be less effective near the poles than near the equator as a result of the meridional cooling discussed above. Danielson assumed that there was no meridional cooling.

It is not clear that Danielson's model is entirely consistent, even within the context of his calculations. According to this model, the amount of methane in the atmosphere is controlled by the temperature of the summer pole, which he finds to be 80°K. The corresponding abundance of methane is 2 km-A. However, as discussed above, the observed abundance of 2 km-A may be significantly less than the total atmospheric content of methane, since most of the line formation takes place near a discrete cloud layer. Furthermore, Danielson assumes the dominant methane surface ice is pure methane, while Lewis suggests that methane clathrate is more likely. The vapor pressure curve of methane clathrate is orders of magnitude below that of solid methane. Danielson's model would be inconsistent with a surface containing methane clathrate.

Conclusions

Titan has a warm upper atmosphere. Whether it has a warm lower atmosphere and hot surface is an open question at present, which future radio and infrared observations from the ground and from aircraft can help answer. This question is important not only from a purely scientific point of view, but also for assessing the viability of a Titan Entry Probe. As current entry probes would begin to make measurements at about the 7 mb level, the value of Titan's surface pressure or at least a lower bound is of extreme importance.

Sagan: The ammonia curve in Figure 2-28 is far below an observed broad-band point. Is this consistent with your suggestion of ammonia in the lower atmosphere?

Pollack: The new narrow-band points are far below this broad-band point. This perhaps indicates the presence of another emission line, as suggested by Danielson and Caldwell.

Veverka: From the measured infrared spectra, can you exclude the presence of large amounts of helium?

Pollack: Unfortunately, no.

Danielson: Your greenhouse models require large amounts of methane, on the order of 40 km-A. With such gas amounts, there will be a significant amount of Rayleigh scattering at short wavelengths and a lot of absorption at longer wavelengths by gaseous methane. In this sense, is your model consistent, that is, will enough sunlight reach the surface to cause a greenhouse effect?

Pollack: As I discussed above, it is even more difficult to have a greenhouse work in the case of Venus. Yet, the Venera-8 results imply that it is operative for Venus. A greenhouse can work as long as a minimal amount of sunlight reaches the surface. The most serious constraint on a greenhouse of Titan is the requirement that the atmosphere is massive enough, that is, that the surface pressure exceeds 0.1 atm.

Danielson: The situation for Titan may be even worse than for Venus in view of the low albedo of the clouds.

Pollack: Not really. If the clouds have a low albedo because they have an optical depth not much above unity, then more light will be transmitted through the clouds on Titan. Furthermore, the surface pressures I consider for Titan are orders of magnitude lower than for Venus. When Boese finishes his important laboratory study of methane at low temperatures, I hope to construct some greenhouse models for Titan that allow for solar energy deposition in the atmosphere.

Veverka: In performing the circulation calculations, you made two assumptions that are reasonable, but we are not sure they are true. One is that Titan has a synchronous rotation period and the other is that its rotation pole lies perpendicular to the orbital plane.

Pollack: I would be very surprised if Titan did not have a synchronous period, since we know that other less massive satellites of Saturn do.

Hunten: Also, the Lewis models of the interior of Titan imply that it is highly dissipative.

Danielson: In the case of a transparent atmosphere, can frictional heating by winds at the surface boundary help equalize the surface temperatures?

Pollack: Based on my experience with Mars, I am very skeptical that this would be important. I believe the key factor is radiative exchange between the atmosphere and surface, which is important for your model as well as the greenhouse model.

Trafton: If there is a high altitude cloud, then its bottom would have to be located in a zone of changing temperature.

Pollack: As you mentioned in your talk, your observations appear to imply a discrete cloud layer. My methane cloud would fit that requirement, and it is located in a region where the temperature is still noticeably decreasing in my model atmospheres. I might mention that it is not at all clear how one would get such a discrete cloud within the context of Danielson's model.

Hunten: Sagan's photochemically produced aerosols will eventually collect as a deposit of brown polymer on the surface. It might be as much as a few kilometers deep. This layer could prevent contact between gaseous methane in the atmosphere and solid methane clathrate and hence vitiate some of the above vapor pressure arguments.

2.7 ULTRAVIOLET OBSERVATIONS OF TITAN FROM OAO-2

J. J. Caldwell

Introduction

I would like to briefly discuss some broad-band photometry measurements from OAO-2 and how they reflect on atmospheric models of Titan. These data extend the observed reflectivity of Titan down to 2600 Å and thus further clarify geometric albedo variations in the ultraviolet.

Observations and Interpretation

Figure 2-31 summarizes several sets of measurements of the reflectivity of Titan in the wavelength range 0.26 - 1.1 μm. In deriving geometric albedos from broad-band photometry data, both Harris (1961) and Caldwell (1973) used observations of solar-type stars to effect the necessary division of the Titan data by the Sun, rather than using an absolute calibration. Instrumental sensitivity is effectively canceled by this technique. In the reduction of the OAO-2 photometry, Titan's radius was taken to be 2425 km (Dollfus 1970) and the solar visual magnitude (V_o) was taken to be -26.74 (Johnson 1965). The geometric albedos of Harris were modified slightly to incorporate these more modern fundamental data. No correction was made for the slightly different solar colors used in the two analyses: the effect of such a correction would have been to raise the Harris albedos slightly relative to the OAO-2 points. Also shown are the relative reflectivity measurements of McCord et al. (1971), scaled to match the other data in the blue and ultraviolet. The OAO-2 results are summarized numerically in Table 2-6, and UBVRI colors in Table 2-7.

For the purpose of this discussion, the most important point to note is the extreme drop in albedo toward the ultraviolet. The measurements of Barker and Trafton (1973) (see also Section 2.3) are in very good agreement with the present results. The OAO-2 data extend the wavelength range slightly over the ground-based range, and emphasize that the albedo does not begin to turn up again, at least for wavelengths longer than 2600 Å. (The shortest wavelength data point originally published by McCord et al. at 3000 Å, is considered to be unreliable, and has been suppressed.)

It has been emphasized elsewhere (e.g., Veverka, Section 2.4) that this trend in the albedo is inconsistent with the expected Rayleigh scattering of the amount of a molecular atmosphere invoked by Trafton (1972) to explain Titan's spectroscopic properties. For example, note the ultraviolet Rayleigh scattering of 2 km-atm of CH_4 shown in Figure 2-31. It may be concluded that there is an additional absorbing constituent in the ultraviolet. Furthermore, this absorption must occur high in the atmosphere, if it is to mask the Rayleigh scattering.

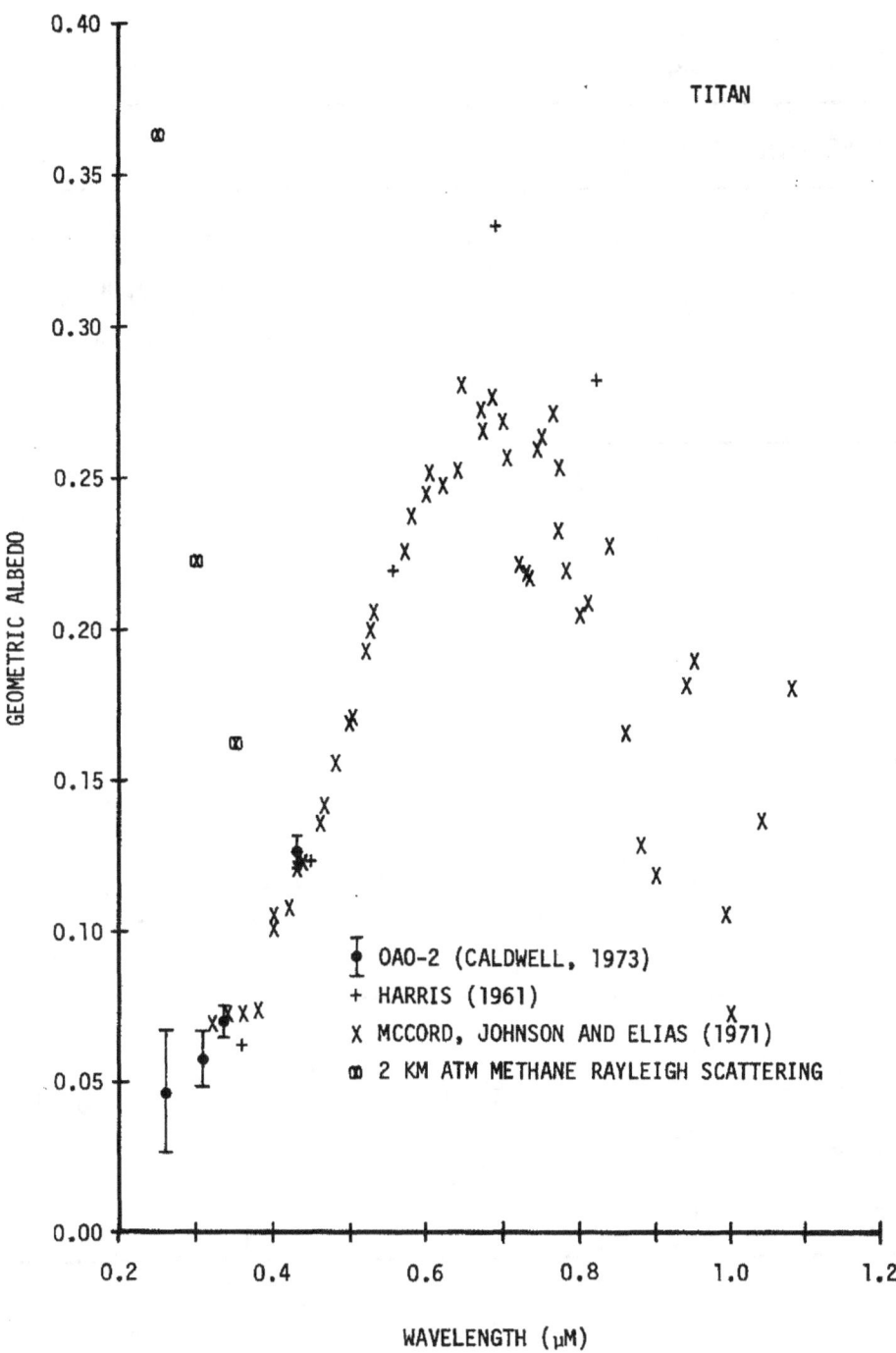

Figure 2-31. Geometric albedos derived from broad-band photometry from OAO-2,
together with ground-based data by Harris (1961). Harris' data
have been adjusted to $V_O = -26.74$ (Johnson, 1965) and Titan
radius = 2425 km (Dollfus, 1970). The relative reflectivity
determined by McCord et al. (1971) has been scaled to the other
blue and ultraviolet measurements. Also shown is the calculated
geometric albedo for 2 km-atm of CH_4.

Table 2-6. OAO-2 Titan Ultraviolet Geometric Albedos

EFFECTIVE WAVELENGTH (Å)	ALBEDOS
4300	0.126 ± 0.006
3360	0.070 ± 0.005
3075	0.057 ± 0.001
2590	0.047 ± 0.018

Table 2-7. Photometric Colors

	TITAN	SUN
U-B	0.75	0.21
B-V	1.30	0.65
V-R	0.88	--
R-1	0.11	--

Summary

High altitude deposition of energy in Titan's atmosphere can have a significant effect on the spectral distribution of emitted thermal radiation from the satellite. This reasoning led to the prediction (Caldwell et al. 1973) of emission peaks at wavelengths corresponding to allowed bands of CH_4 (7.7 μm) and trace photolysis products such as C_2H_6 (12.2 μm). The subsequent publication (Gillett et al. 1973) of intermediate resolution infrared spectrophotometry has encouraged this interpretation of the infrared properties of Titan, and provided the basis for an initial, detailed model, presented in the following paper by Danielson et al.

2.8 AN INVERSION IN THE ATMOSPHERE OF TITAN*

R. E. Danielson, J. J. Caldwell, and D. R. Larach

Introduction

Measurements of unexpectedly high infrared brightness temperatures on Titan in the 8-14 micron window (Low 1965, Allen and Murdock 1971, Gillett, Forrest, and Merrill 1973) and a lower temperature in the 20-micron window (Morrison, Cruikshank, and Murphy 1972) have been widely interpreted in terms of greenhouse models (Pollack 1973, Sagan 1973) where the 8-14 micron radiation originates at or near the surface while the 20-micron flux is emitted high in the atmosphere. A very detailed greenhouse model due to Pollack (1973) derives a methane (CH_4) to hydrogen (H_2) ratio of unity (within a factor of 3) and a minimum surface pressure of 0.4 atm. Based on a surface gravity g = 140 cm sec^{-2}, the minimum CH_4 abundance is 30-40 km-A and the minimum H_2 abundance varies from 15 to 85 km-A.

It is the purpose of this paper to propose an alternate model of the atmosphere of Titan which seems to be consistent with observations and requires a much smaller CH_4 abundance (of the order of 2 km-atm). Although no H_2 is required, the presence of some H_2 as reported by Trafton (1972a) is readily accommodated. In this model, a temperature inversion exists in the atmosphere due to absorption of blue and ultraviolet solar radiation by small particles. The absorbed radiation is re-radiated by the dust and by molecules having long wavelength bands such as CH_4 at 7.7 µm and ethane (C_2H_6) at 12.2 µm. The brightness temperature at 20 µm is primarily due to re-radiation by the dust.

The Origin of the Inversion

The continuum geometric albedo of Titan, shown in Figure 2-32, is unusually low in the blue and ultraviolet for an object with an extensive atmosphere. The presence of a large abundance of CH_4 (of the order of 2 km-A) is strongly indicated by the observations of Trafton (1973) which show near-infrared bands of CH_4 having widths comparable with those in Uranus. The observed geometric albedo p = 0.05 ± 0.02 at 2600 Å (Caldwell 1973) can be produced by the Rayleigh scattering of about 0.15 km-A CH_4 overlying a black surface. Under the same conditions, 1 km-A CH_4 will produce p = 0.22 and 10 km-A will yield p = 0.60. One must therefore conclude that some substance is strongly absorbing in the blue and ultraviolet. We propose that this absorption is due to small particles (hereafter called dust) produced as a result of photolysis in the uppermost portions of the atmosphere. A dark reddish-brown polymer with properties qualitatively similar to those required of the dust has been produced in the laboratory by Khare and Sagan (1973).

If these dust particles were very large, they would radiate like black bodies with emissivity, $\varepsilon \approx 1$, and their temperature would approach 90°K, the black sphere temperature at the distance of Titan from the Sun. On the other

---

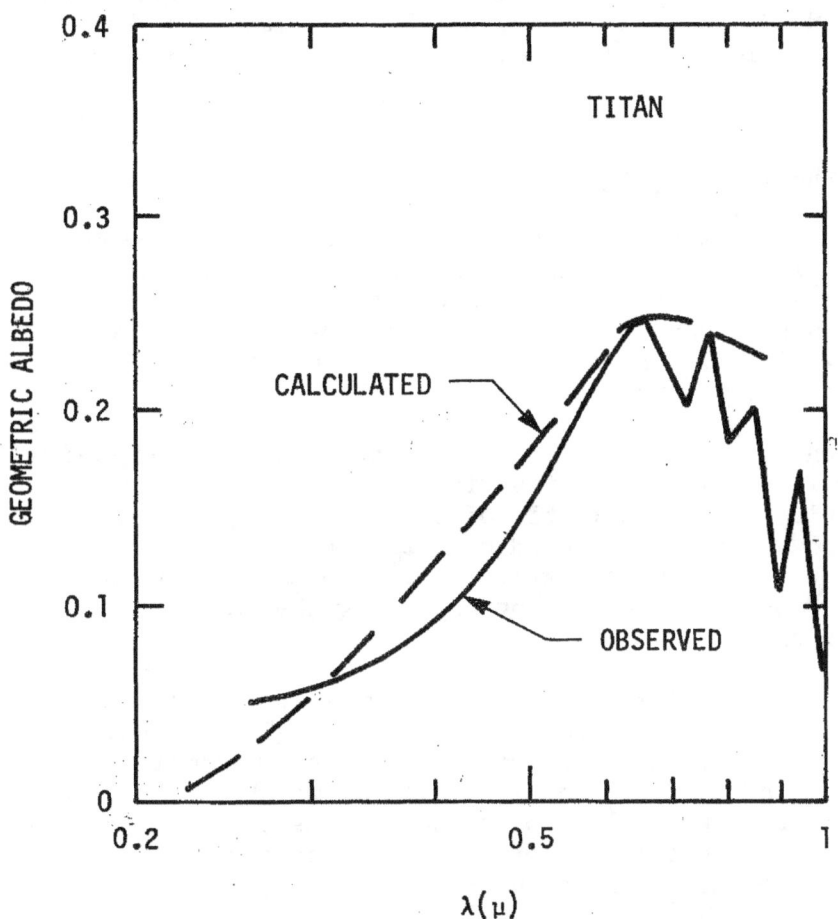

Figure 2-32. The observed geometric albedo is based on the observations of
McCord, Johnson, and Elias (1971) normalized to p = 0.20 at
λ = 0.555 μm and on the OAO observations of Caldwell (1973).
The calculated curve is from the simplified model in Section 4.

hand, if the particles are small compared with the wavelengths characteristic of a 90°K Planck spectrum, they will be poor emitters and will rise to temperatures higher than 90°K. The gaseous atmosphere will be heated to nearly the same temperature as the dust due to collisions of gas molecules with the dust particles. The final temperature will be determined by the emissivity of the dust and the available molecular bands. Methane has no allowed bands longward of 7.7 μm and hydrogen, which has no allowed dipole bands, has negligible opacity due to collision induced transitions at the atmospheric pressures characteristic of our model (of the order of 0.01 atmosphere). Among likely additional constituents, the longest allowed bands are 10.5 μm (ethylene, C_2H_4), 12.2 μm (ethane, C_2H_6) and 13.7 μm (acetylene, C_2H_2). No plausible molecule has any appreciable emission longward of 15 μm. If such a molecule were present, no large inversion could be sustained.

The Emission of the Titan Atmosphere

To estimate the emission of the Titan atmosphere, we assume (for simplicity) it is isothermal except in the boundary layer near to the surface. Some justification for this assumption is given in Section 6. We fix the atmospheric temperature at 160°K by noting that the brightness temperature measured by Gillett, Forrest, and Merrill (1973) at 8.0 μm, which is near the center of the very strong CH_4 band at 7.7 μm, is nearly 160°K. (See Figure 2-33.)

The emission of the dust is calculated to be that of an optically thin medium radiating at 160°K. The emissivity of the atmosphere is assumed to vary as λ^{-1}, which would be characteristic of particles which are small compared with the wavelength and which are composed of a substance whose complex index of refraction is independent of wavelength. The curve of dust emission shown in Figures 2-33 and 2-34 is calculated by adjusting the emissivity to agree with observations at 9 and 10 μm.

We propose that the large peak near 12 μm is due to the 12.2-micron band of C_2H_6. Based on a band strength of 24 cm^{-1}/cm-A (Thorndike 1947), we estimate that the amount of C_2H_6 required to produce a mean emissivity of ∿0.1 at the 12.2-micron peak is of the order of 1 cm-A. The width of the 12-micron feature as shown in Figure 2-34 was taken to be the same as a similar feature at 12 μm observed in Saturn (Gillett 1973, private communication). Laboratory measurements and detailed calculations on the 12.2-micron C_2H_6 band will be required to establish its width under the conditions in the Titan atmosphere.

The Energy Balance of Titan's Atmosphere

The energy balance of this model of Titan will be illustrated by an idealized calculation in which Rayleigh scattering is ignored for simplicity. Adopting a radius of 2550 km, as did Morrison, Cruikshank, and Murphy (1972) based on a correction to the measurements of Dollfus (1970), the visual geometric albedo (p) equals 0.20 if the visual magnitude of Titan at mean opposition is taken to be 8.39 (Harris 1961) and if the absolute visual magnitude of the Sun equals -26.78 (Allen 1963). Using the observations of spectral reflectivity given by McCord, Johnson, and Elias (1971), one obtains the curve of observed geometric albedo shown in Figure 2-32. The maximum geometric albedo (p ≈ 0.25) occurs at $\lambda \approx 0.65$ μm, or approximately the same wavelength as the minimum absorption of the brown polymer (dust) shown in Figure 2-35.

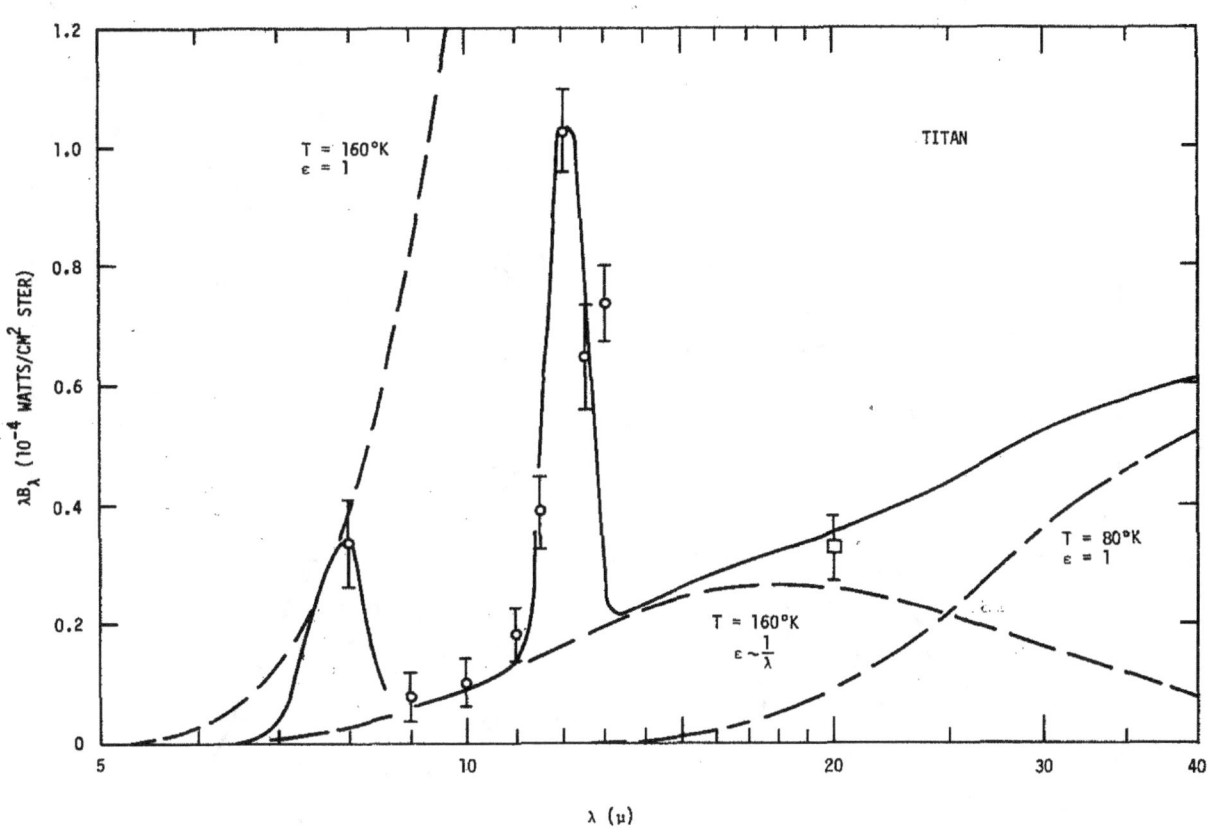

Figure 2-33. The solid curve shows the predicted emission spectrum based on
a highly simplified inversion model of Titan. The peaks at
7.7 μm and 12.2 μm are due to emission by CH_4 and C_2H_6 bands.
The emission by the dust is shown as a 160°K black body having
an emissivity which is inversely proportional to the wavelength.
The radiation from the surface is shown as an 80°K black body.
This graph is constructed in such a way that the area under
the curves is proportional to the energy radiated. In both
this figure and in Figure 2-34, the filled circles are data
by Gillett et al. (1973), and the filled square is a measurement by Morrison et al. (1972).

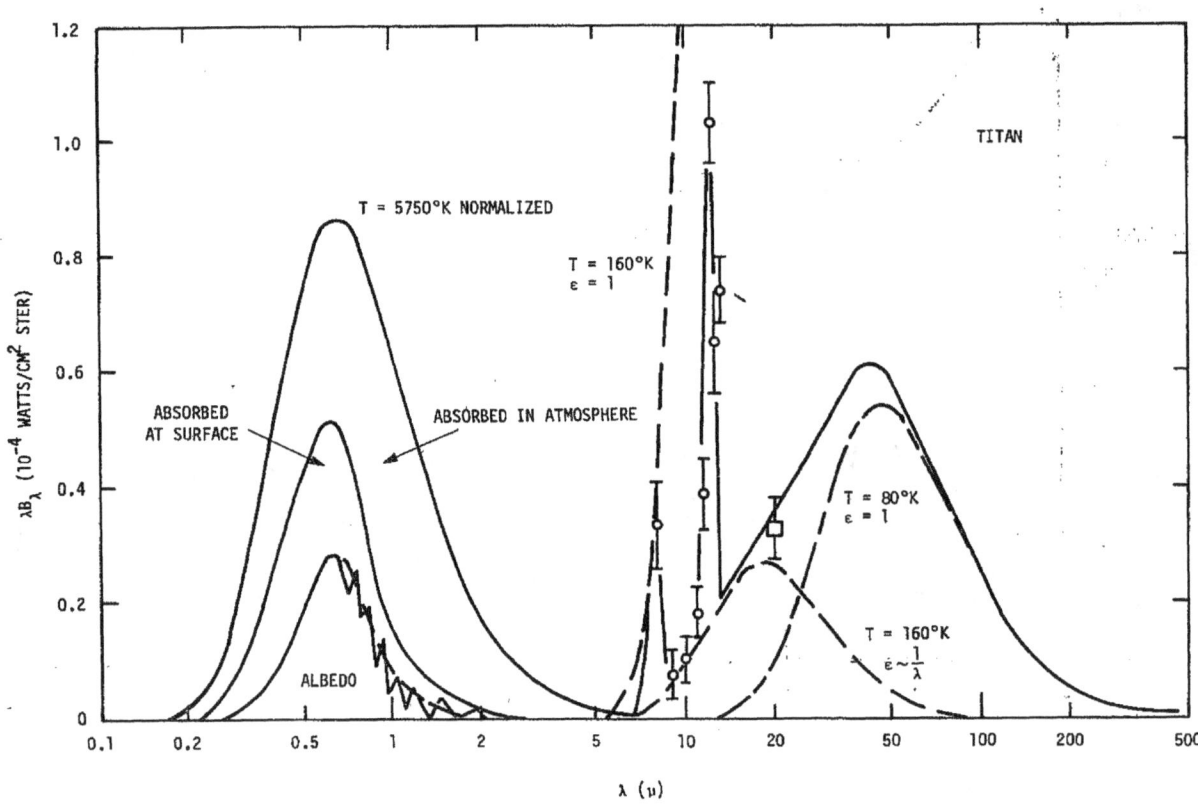

Figure 2-34. The energy balance of the inversion model of Titan is illustrated
in this figure where area is proportional to energy. The area
under the curve representing the incident solar radiation (T =
5750°K) is equal to the area under a 90°K black body curve. Most
of the solar radiation is absorbed by the dust in the atmosphere
and re-radiated in the far infrared.

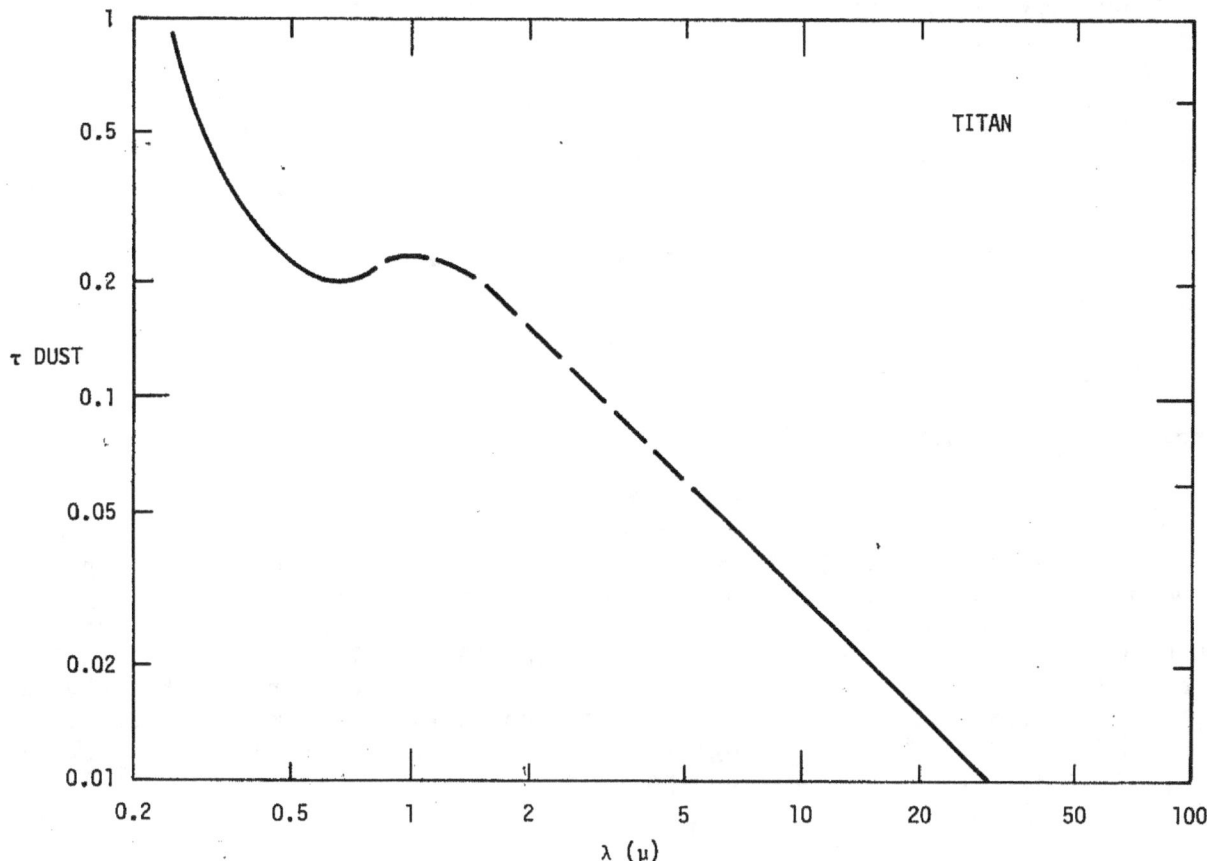

Figure 2-35. The optical thickness of the dust suspended in the Titan atmosphere. The solid curve shortward of 0.8 μm is based on the measurements of Khare and Sagan (1973) normalized to agree with the observed geometric albedo. The curve longward of 5 μm is based on an assumed λ^{-1} dependence normalized to agree with the observed flux at 9 and 10 μm.

In order to have a definite exploratory model, the surface albedo of Titan is taken to be proportional to wavelength up to λ = 0.65 μm, above which it is assumed to be 0.65. The rationale behind this arbitrary choice of surface albedo comes from the working hypothesis that the surface of Titan is covered by snow whose reflectivity is governed by the dust which has settled out of the atmosphere.

The geometric albedo, p, and Bond albedo, A, of a purely absorbing atmosphere (obeying Beer's law) which has an optical depth, τ, overlying a Lambert surface having a reflectivity R can be shown to be:

$$p = \frac{2}{3} Re^{-2\tau} [1 - \tau + 2\tau^2 - 4\tau^3 e^{2\tau} E(2\tau)], \tag{1}$$

and

$$A = Re^{-2\tau} [1 - \tau + \tau^2 e^{\tau} E(\tau)]^2, \tag{2}$$

where:

$$E(y) = \int_y^\infty \frac{e^{-x}}{x} dx. \tag{3}$$

From Equation 2, p = 0.25 yields τ = 0.20 which establishes the normalization of the visible and ultraviolet portions of the curve of τ_D, the optical depth of the dust in the atmosphere, shown in Figure 2-35. Figure 2-35 is based on the measurements of the transmission of a thin layer (thickness of the order of 0.03 mm) of brown polymer by Khare and Sagan (1973). It can be shown that the optical depth of a layer containing particles (small compared with the wavelength) is approximately equal to that of a slab having the same mass per unit area if k << 1, where k is the imaginary part of the index of refraction. This condition is satisfied for the brown polymer, for which k is the order of 3 x 10^{-3}.

The adopted variations of τ_D and R lead to the calculated curve of p shown in Figure 2-32. For the purposes of this paper, the predicted geometric albedo is in satisfactory agreement with that observed shortward of 0.65 μm. The Bond albedo shown in Figure 2-34 is based on the crude assumption that Beer's law holds for the CH_4 absorption longward of 0.65 μm. The albedo for the incident solar radiation is 16%; the corresponding black sphere temperature is 86°K.

Under the same assumptions governing equations 1-3, the fraction of the intercepted solar radiation which is absorbed at the surface is given by:

$$f_s = (1 - R)e^{-\tau} [1 - \tau + \tau^2 e^{\tau} E(\tau)] \tag{4}$$

Based on Equation 4, 23% of the solar radiation is absorbed at the surface. The remaining 61% is absorbed in the atmosphere. Since the atmosphere is optically thin in the far infrared (except near 7.7 μm), approximately half of the absorbed radiation is reemitted downward and is absorbed by the surface. This absorbed radiation would, by itself, keep the surface at a temperature of 67°K.

Absorbed radiation: 1.13 x 10^{-4} $\frac{watts}{cm^2}$

98

From Figure 2-34, the radiation emitted (upward) by the atmosphere equals 1.38×10^{-4} watts/cm^2 which is 22% larger than half of the radiation absorbed in the atmosphere. In a fully self-consistent model, these numbers would be equal. (They would be equal if the far infrared emissivity of the dust decreased a little faster than λ^{-1}, for example.) <u>The important conclusion to be drawn from the above discussion is that the solar radiation absorbed by the dust is sufficient to maintain a large inversion (temperature approximately 160°K) in the entire atmosphere of Titan.</u>

From the emitted flux and the heat capacity of the atmosphere, one may compute its rate of cooling when it is not illuminated by the Sun. The result is about 0.1°K/day if the atmosphere contains 2 km-A of CH$_4$. Thus the change in atmospheric temperature is insignificant during Titan's solar day (16 days if its rotation is synchronous with its orbital period). Furthermore, the cooling rate is sufficiently slow to allow the atmospheric temperature over the winter pole to be maintained near 160°K (except near the surface) by lateral transport (winds).

The Surface Temperature of Titan

If the atmospheric radiation were the only source of heating the winter poles (which are inclined 27° if the axis of Titan is parallel to the axis of Saturn), the surface temperature would only be 67°K. The vapor pressure of solid methane at this temperature is about 10^{-3} atm corresponding to a methane abundance of about 0.1 km-A. This abundance is too small to explain the width of the Titan methane bands observed by Trafton (1973). If the minimum surface temperature is 80°K, the corresponding CH$_4$ abundance is 2 km-A and the C$_2$H$_6$ abundance is of the order of 5 cm-A. Extending the calculations in Section 4, we find that the surface temperature at the sub-solar point would rise to nearly 100°K assuming the surface has negligible thermal inertia. Under the same assumptions, Figure 2-36 shows the variation in surface temperature on the illuminated portion of the surface.

Undoubtedly some heating of the unilluminated portion occurs as a result of surface winds driven by surface temperature differences. In our model, however, the main source of heating (in addition to the atmospheric radiation) is by means of the latent heat released as a result of the condensation of solid methane. We propose that the winter polar regions are kept near 80°K as a result of the condensation of CH$_4$ gas. Reduction of the albedo of the polar caps by the dust which has settled out is the vital factor which enables the condensed snows to be resublimed in the polar summer. In the absence of the dust, the summer pole would be much too cold.

During the 30-year long Titan year, the surface pressure may vary somewhat with time because the sublimation and condensation would not always be equal. We can estimate the magnitude of this variation by computing the decrease in CH$_4$ abundance required to keep a polar cap (down to a latitude of 63°) at a temperature of 80°K for 15 years. The result is 0.08 km-A of CH$_4$ which is small compared with the abundance of 2 km-A adopted in our exploratory model.

99

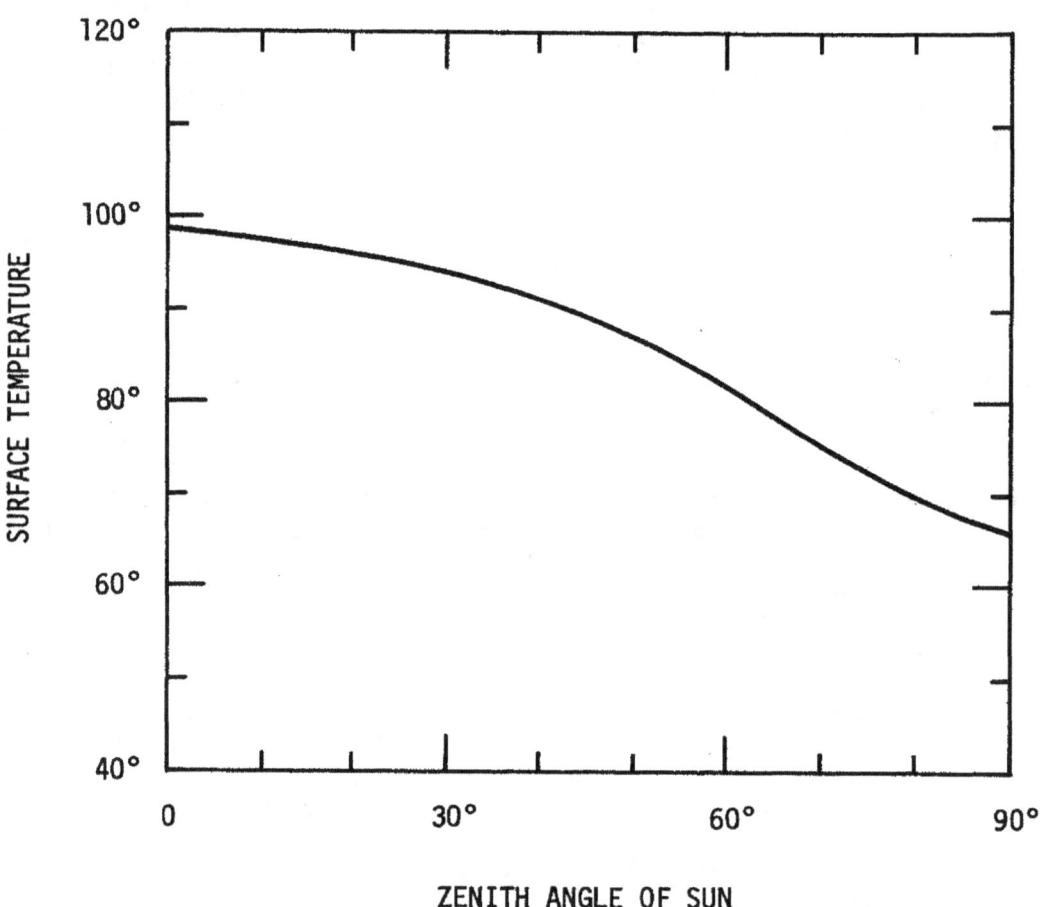

Figure 2-36. The surface temperature of Titan predicted by ignoring the thermal
inertia of the surface, the energy transfer by surface winds, and
the latent heat of CH₄. Inclusion of the latent heat is expected
to make the surface temperature more uniform than shown.

Similarly, methane condensation will supply heat to the equatorial regions at night and sublimation will occur during the daytime. Much, if not all, of the surface of Titan should be covered with frozen CH_4 and other ices. If this be the case, the surface temperature will be more uniform than indicated in Figure 2-36 due to the thermostatic effect of the solid CH_4. In Figures 2-33 and 2-34, the predicted radiation from the surface is indicated as an 80°K black body.

Summary Discussions of the Model

With the aim of introducing the basic ideas as clearly as possible, the inversion model of Titan presented in this paper is highly simplified, thereby avoiding prematurely complex calculations. If the model continues to be viable after further scrutiny, many of its features should be studied in more realistic detail.

A more detailed analysis must include the effects of Rayleigh scattering. We have made some calculations including scattering which indicate that the visible and ultraviolet optical depth of the dust must increase more rapidly than shown in Figure 2-35 in order to explain the observed geometric albedo. Macy (1973) finds the same to be true for Saturn. It therefore seems that the dust produced in planetary atmospheres exhibits a more rapid increase in absorption toward the ultraviolet than does the substance produced by Khare and Sagan (1973). Further laboratory investigations in which mixtures of only CH_4 and H_2 are irradiated with hard ultraviolet photons may be required. The full index of refraction (real and imaginary parts) of the substances produced should be measured.

We believe that detailed knowledge of the properties of the dust will be important in understanding other planets as well as Titan. Both Jupiter and Saturn exhibit absorption features which begin in the red and increase toward the blue. Preliminary attempts to explain these features in terms of absorption by dust particles similar to those postulated for Titan have been made for Jupiter (Axel 1972) and for Saturn (Macy 1973). Uranus may also have some absorbing particles in its atmosphere (Light and Danielson 1973). Hence the major planets should have substantial inversions in the upper portions of their atmospheres where the density is $\approx 10^{18}$ cm^{-3} or less. Such an inversion has been observed on Jupiter (Gillett, Low, and Stein 1969, Gillett and Westphal 1973). Indeed, if Titan should have a massive atmosphere of the type proposed by Pollack (1973), an inversion would occur in the upper portion of its atmosphere.

We are of the opinion that the dust particles are composed mainly of higher hydrocarbons from the by-products of the photolysis of CH_4 by solar ultraviolet photons having wavelengths shorter than 1600 Å. Strobel (1973, see also Strobel and Smith 1973) has performed detailed studies of the photochemistry of hydrocarbons in the Jovian atmosphere. He finds that approximately 20% of the dissociated methane is irreversibly converted to higher hydrocarbons. If a similar percentage is valid for Titan, of the order of 10 km-A of CH_4 should have been converted over the lifetime of the satellite. Strobel's (1973) calculations suggest that C_2H_6 may be the main substance formed. In our inversion model, almost all of the C_2H_6 will freeze out on the surface leaving only a few cm-A of gaseous C_2H_6 in the atmosphere.

Our most recent attempt to model the emission peak at 12.2 μm is shown in Figure 2-37. Using the formulae for a symmetric top molecule given by Herzberg (1950) and random band transmission formulae from Goody (1964), the disk-integrated thermal emission from an isothermal (160°K) atmosphere of 2 km-atm of CH_4 and 0.5 cm-atm C_2H_6 has been calculated. Also shown are some of the data of Gillett et al. (1973). The agreement is fair near the center of the band, but breaks down away from the center. This may indicate that the molecular model used is not correct, or there may be additional trace materials present. The discrepancy between the 0.5 cm-atm of C_2H_6 in this calculation, and the 5.0 cm-atm quoted earlier, could result from the uncertain extrapolation of measured vapor pressure data points to low temperature, or perhaps partly from the Titan inversion not extending down to the surface.

Based on a surface temperature of 80°K, the corresponding abundance for C_2H_2 is $\sim 10^{-2}$ cm-A and ~ 40 cm-A for C_2H_4. The latter abundance is sufficiently large that the atmosphere should be optically thick at 10.5 μm (the band strength is about 500 cm^{-1}/cm-A) yielding a brightness temperature of about 160°K. Although no narrow-band measurement has been made at 10.5 μm, the measured flux from 10-12 μm is consistent with a substantial emission peak at 10.5 μm (Gillett, Forrest, and Merrill 1973). In addition, the observed brightness temperature at 13 μm suggests the possibility of some atmospheric emission at 13.7 μm due to C_2H_2. These two possible emission peaks were not included in our exploratory model.

A more detailed analysis of this model should also include a calculation of the vertical temperature distribution. Since the grains reradiate most of the radiation they absorb, the atmospheric temperature is mainly determined by the solar radiation at each elevation above the surface. The decrease of incident solar radiation with depth in the atmosphere is somewhat compensated by the increased amount of outgoing solar radiation (reflected by the surface) in the lower atmosphere. Hence the assumption of an isothermal atmosphere may not be too unrealistic.

Acknowledgement

This work was supported by NSF Grants GP-23580 and GP-39055. We are grateful to Dr. F. Gillett and to Dr. B. Khare and Dr. C. Sagan for providing their data in advance of publication.

Morrison: I would like to raise a question that perhaps several people would want to address. A number of you talk about models in which all, or a significant fraction, of the radiation has to reach the surface. You've said this about the poles and Pollack's models, as I recall, assume that all radiation is deposited at the surface. But Veverka has talked about optically thick clouds. I wonder what both of those things mean. In an optically thick cloud, what fraction of the total radiation can get through to the surface? When you say you assume that large fractions, or all, of the radiation reaches the surface, what really do you mean are the limitations?

Danielson: Well, we had a luncheon discussion on that very subject, and if I may, let me summarize it for you. What Veverka says is that a dense cloud is needed. An alternate version of that would be a snow field with the absorbing dust in the atmosphere. Veverka agrees that a snow field will give him the

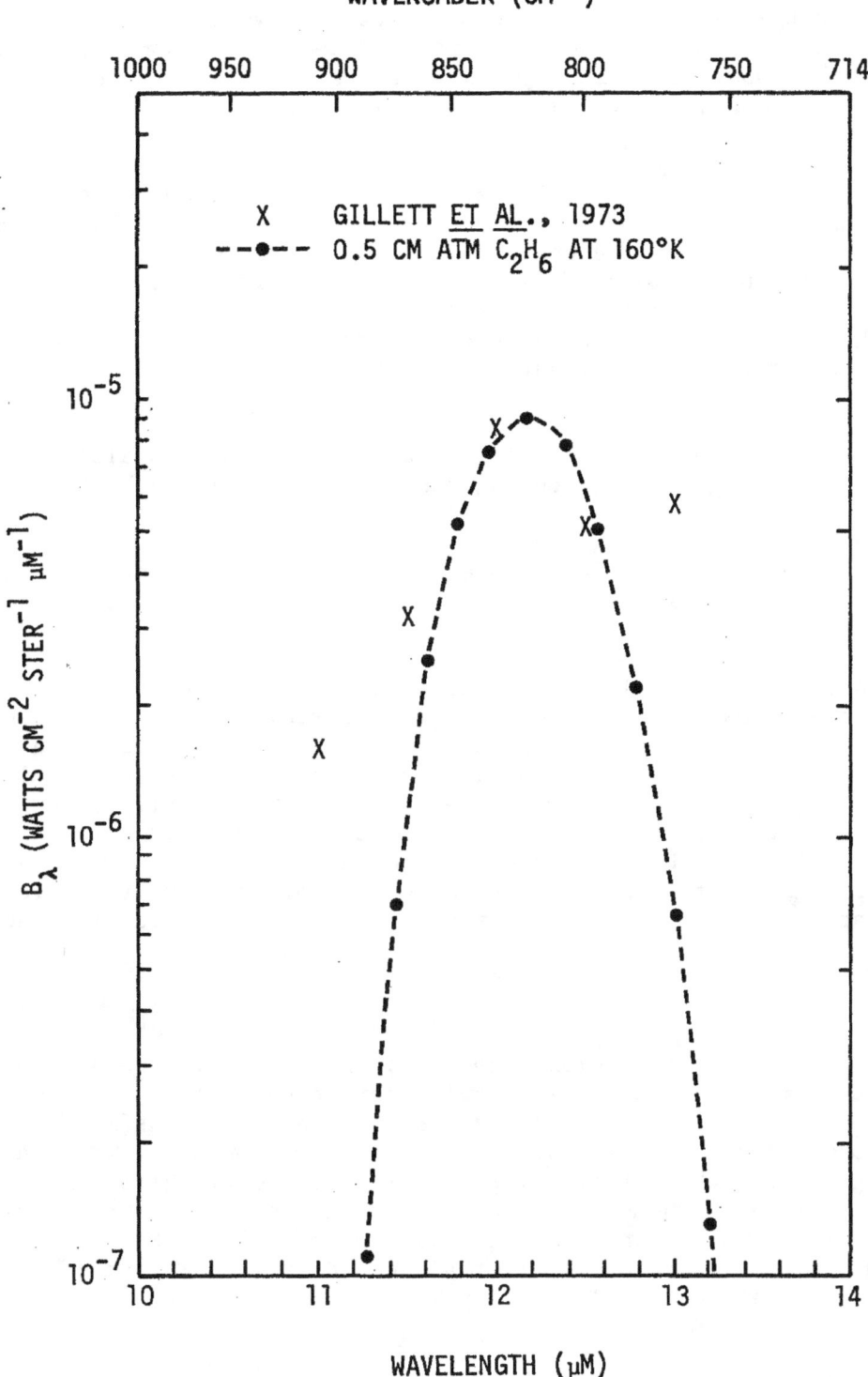

Figure 2-37. The thermal emission from an atmospheric inversion consisting of a 160°K atmosphere (2 km-atm CH₄, 0.5 cm-atm C₂H₆) above an 80°K surface. The emission is averaged over a hemisphere.

variations of polarization with phase that he needs, but it has to be bright, and we know that Titan is not bright. However, if we put in these nonscattering particles which are purely absorbing, it is like putting a neutral density filter in front of the snow. Then it might have the same polarimetric properties for all we know at the moment; isn't that right, Joe (Veverka)?

Veverka: In principle, yes. It is certainly true that, out of hand, one cannot reject your model on the basis of the phase coefficients or the polarization measurements. What has to be done, of course, is to perform the calculations and see if that kind of model fits. But, I would also like to address the question of getting energy through clouds to the surface. This is the problem with Venus, right? You have optically thick clouds and you have no problem getting energy to the surface, or at least quite deep anyway.

Danielson: Well, Venus' clouds are optically thick but the albedo is high, so you have τ^{-1} reaching the surface in conservative isotropic scattering.

Morrison: Could Pollack or Sagan make an elucidating comment as to how this affects their models? I'm not sure I understood what was just said about Venus.

Pollack: Well, let me explain the Venus story in a somewhat different way. That is, that a greenhouse model, in order to work, really doesn't need very much solar energy reaching the surface. It needs some nonzero amount to reach the surface, but that nonzero amount, as in the case of Venus, could be very small. Now, in the case of Titan, I think the real question is how optically thick are the clouds. To satisfy observational measurements, like the polarization, I'm sure doesn't require huge optical thickness as the lower bounds. To just throw a number out, I would rather suspect that optical depths on the order of unity would satisfy all the observational constraints, like the polarization. Certainly under those conditions a significant amount of radiation will reach the surface. Now, obviously if it's a lot higher it's a whole new ball game.

Danielson: I would like to emphasize that, in our model, the atmosphere has no clouds, and I am worried about Trafton's contention that you must have clouds to fill in these methane bands. Frankly I don't see how clouds in the upper portion of the atmosphere can form in the presence of our proposed inversion. An action item I am going to take for myself is to try and see if there isn't some way to get around Trafton's argument on the clouds. It is the most difficult one that I am aware of, at the moment, for this atmosphere which has no clouds unless you want to call the dust a cloud. Veverka's cloud is a snow covered surface, but there's no way that that would satisfy Trafton.

Trafton: But my cloud need not be at the same altitude as your inversion. What do you mean by high? You mean optical depth unity in your emission bands, don't you? What do I mean by high? I mean a level above which the methane absorption is negligible compared to the observed methane absorption.

Pollack: I think we're really dealing with methane bands of very different strengths. For example, the 8 μm observation refers to the fundamental of methane where optical depth unity is something like 10^{-3} atm. In the case of Trafton's observations, they are overtone or combination bands that are a lot weaker. So presumably his clouds could be a lot higher and at a pressure level a lot higher than 10^{-3} atm. I think you have to be careful as to where different things are in the atmosphere.

Morrison: Is this also true of the polarization measurement? Could Trafton's clouds be the clouds that one sees photometrically and polarimetrically and all of this exist below your dust layer?

Veverka: Sure. I think I can reverse your argument, Bob (Danielson). I was probably very generous in saying that my observations can be consistent with what you're saying. But I don't see why I can't turn it around. You keep talking about your Lambert surface. Why can't it be a white cloud?

Danielson: Yes, that's quite true. If you wanted a greenhouse model in addition to the inversion, you could replace the snow-covered surface with a cloud. I don't think the cloud would be white though, because of the dust which must be there acting as nucleating particles for it.

Trafton: What about the emissivity of this dust? It seems as if your whole model is predicated on this emissivity being so low that the temperature of the dust is disproportionately high, and this in turn results in the mechanism for your model of a temperature inversion. Basically, it's the non-unit emissivity of the dust which allows you to construct your model. What is the observational evidence that dust particles, small compared to the wavelength of light, are higher in temperature than their surroundings?

Danielson: There is no observational evidence. It is purely a theoretical argument but I think it is a pretty strong one. Think of the particle as an antenna. Just try to radiate at one meter with such a small antenna and you find that it's a very inefficient antenna. That's the origin of this λ^{-1} dependence for small particles.

Trafton: You mentioned this is observed in interstellar dust?

Danielson: Yes, isn't it true that the interstellar dust, if it were simply in equilibrium with starlight and such, it would be 3°K, whereas typical temperatures are believed to be around 100°K.

Trafton: Well, contrary to the situation with the interstellar medium, in Titan's atmosphere the dust particles are going to be colliding with a lot of methane molecules and stuff like that. How do you know that the collisional interactions here won't be enough to cool down the dust?

Danielson: The dust and the molecules go to the same temperature. The molecules have limited ability to radiate it away since they can only radiate in the 12-micron and 7.7-micron bands. If you could invent another molecule that was a good strong radiator at 30 μm, then the dust would cool down and there would be no inversion. But as long as you don't have any emissions longward of 15 μm, I believe my argument is correct, as long as the particles are small. That is, small compared to the 40-micron wavelengths, so 1 μm or half a micron particles would be quite adequate for that.

Sagan: In other words, things that radiate at 30 μm, which is the pure rotation spectrum, like CH_4 and H_2, aren't awfully good for that, but what about any molecule that's not symmetric?

Danielson: Well, name one. That's the problem.

Sagan: Asphalt.

Danielson: That becomes a solid, forming my little particles. The vapor pressure very quickly eliminates almost all molecules. That's the trouble we found at first when we were trying to fill up this 8-14 micron region with emitters. Everything solidifies very quickly, and ethane and ethylene are really the only two effective constituents I'm aware of. Caldwell's question on silane was motivated by this as a substance whose vapor pressure is substantial. We wondered if that might play a role because it has an emission somewhere in this region of 9-12 μm. However, unless we can short-circuit some of Lewis' comments, it's also hard to get.

Trafton: There's one other point about the dust I'm concerned about. You might expect the dust to have a distribution rather than coming out something like all the same size, and, therefore, a certain fraction would be larger than the averag. Since a lot more flux from the exponential side of the emission function occurs a longer wavelengths, maybe this could compensate, or more than compensate, to the point where you could have significant long wavelength radiation from the particles.

Danielson: Well, you've got to have something like a 15 μm diameter particle, and that's going to drop out of the atmosphere awfully quick.

Sagan: Let me just say a word on that. We have measured the particle sizes of our polymers that you are using here. Now I'm sure that the particle sizes are dependent on some conditions which may not be the same as on Titan, but, to whatever extent you want to use our models, the average particle size is 100 μm.

Danielson: They won't stay in the atmosphere. They'll go plop.

Sagan: They will go plop, but there will be others that'll be made. The question is: might the cloud be a steady-state concentration between (1) formation, (2) growth to 100 μm, and (3) Stokes-Cunningham fallout? The answer to that might be yes. If the answer is yes, and you have particles much larger than the infrared emission wavelengths, then, is it true that you are in trouble trying to stay hotter than the equilibrium temperature?

Danielson: I think so. That would do the trick.

Hunten: How many solar constants of UV do you have in your experiments, Carl (Sagan)? Grossly, you'd expect the growth rate of the particles to be proportional to the UV flux, other things being equal.

Sagan: But surely, the particle size depends on the fall-out rate as well as the UV flux.

Hunten: Yes, radiation, the flux, the time, the rate of fallout; and obviously fallout is a lot slower on Titan than in your flask. It seems to me very difficult to estimate particle sizes in Titan's atmosphere on the basis of these lab experiments.

Sagan: Yes, I agree.

Veverka: Do you have a feeling, Bob (Danielson), for how dependent your conclusions are on this λ^{-1} dependence. For example, if the dependence is $\lambda^{-1.5}$ does everything fall apart?

Danielson: Well, actually, it would have been more comfortable in this model to have it $\lambda^{-1.2}$, I believe, because then we would have exact balance between the absorbed and emitted radiation. If it turns out that the complex index varies rapidly with wavelength in a way that is very different than that, then at some point the model is no longer viable. How much it is, I don't know.

Sagan: There's one other thing I wanted to ask about. The C_2 molecules are a central part of the Inversion Model story. They have transitions in the near IR and, particularly, in the near UV. So I wonder what observational limits have been set on these. For example, is there some critical UV observation that could be made, say with the International Ultraviolet Explorer, or from Skylab?

Strobel: The problem with ethane is that it's masked by methane. The onset of absorption for ethane is 100 Å longward of the onset of methane absorption; they almost parallel each other. Since methane is so much more abundant than ethane there's no way you're going to unravel one from the other.

Sagan: Okay, so that's ethane. What about ethylene and acetylene?

Strobel: They absorb at longer wavelengths, out to 2000 Å.

Sagan: So you might have a measurement there.

Hunten: As long as the dust doesn't ruin everything.

Strobel: Yes. There the problem is that as you approach 2000 Å, I think the dust is getting toward an optical depth of one, and all of these constituents absorb below 2000 Å.

Sagan: So the dust does mask the UV.

Danielson: The products formed by the radiation you want mask it. I think the best test I've heard yet is the radio interferometry observations that you mentioned, Carl (Sagan), of the surface temperature. I think that's a crucial thing, and another thing to do is to make really detailed calculations in the 8-14 micron region with the models and just see what fits and what doesn't.

Pollack: I think, though, that you have to be very careful here because there are two questions to be considered which are very distinct from one another. One question is: Is there a significant greenhouse so that the surface temperature is high? Almost independent of the answer to that is the question: What's mainly responsible for the radiation in the 8-14 micron region? It may turn out, for example, that in the whole 8-14 micron region you're just seeing gases and radiation from clouds, and yet have a high surface temperature. That's a plausible model. So, I think we have to be very careful here in separating these two questions.

Sagan: That's a good point. So, determining the surface temperature is not a way of deciding this question on the inversion.

Danielson: That's right. I think, then, that the best way to proceed here is to simply get the best measurements and apply the best theories in the 8-14 μm region.

Hunten: Clearly, a detailed Gillett-type spectrum with points every 0.1 μm would be tremendous. That's possible right now.

Pollack: I think in another half year, Gillett will give us absolute information; so I think in another half year we will be a lot further along.

Morrison: Jim (Pollack), is it really true for the greenhouse models, that you could think up a source of opacity through the 10-micron band such that very little radiation would get out and therefore one could explain the whole 10-micron spectrum by the Danielson approach and still have that compatible with a large greenhouse?

Danielson: Ammonia ... wouldn't that do it?

Pollack: Yes. That is, that the dip at 10 μm is perfectly consistent with the saturated atmosphere, i.e., the lower portion of the atmosphere having ammonia. So I don't think there is any problem there. Even if I didn't want to say that, I could just say, well ... I would have to go away from Danielson's small particles ... but I could say the clouds are optically thick. Thick enough, that is, in the 8-13 micron region that that's all we're seeing there.

Sagan: When I get to my presentation tomorrow, I'll show some substances which are quite opaque at this wavelength. So it's perfectly plausible that the clouds might not let the radiation through.

Hunten: Bob (Danielson), this is really an optical model, not a physical model you put together. You have a medium with dust and gas in it. You don't really have an explicit model for Titan's atmosphere, do you?

Danielson: Well, the temperature is basically determined, at least so far, by the amount of flux that falls on a particle at a given level. I think one could go through this and iterate it once more. If you did, I think you would find that it is not all that variable in temperature until you get right near the surface, where boundary temperature questions cloud the issue.

2.9 BLOWOFF AND ESCAPE OF H_2

D. M. Hunten

Introduction

This discussion is a summary of a recent paper (Hunten 1973a, herein referred to as Paper I) on the escape of hydrogen from Titan. Some extensions and related material will appear in two other papers. Paper II (Hunten 1973b) improves one of the derivations and applies the ideas to several different atmospheres. Paper III (Hunten and Strobel 1974) is a detailed analysis for the Earth, which verifies the general principles discussed here and in Paper I.

Trafton's (1972) announcement of the possible presence of H_2 on Titan included a discussion of its loss by Jeans escape. He did not find a way to specify the height of the critical level, from which this molecular loss can be regarded as occurring. In addition, he implicitly rejected the possibility of hydrodynamic loss, or "blowoff", which can give much greater loss rates under certain conditions. The treatment of Paper I supplies both these missing ingredients. First, it is shown that a pure hydrogen atmosphere cannot be retained by Titan, but will blow off in a few hours. Addition of a heavier gas, such as CH_4 or N_2, in comparable abundance gives a great improvement, although the escape rate can still be large. Moreover, the actual flux can be predicted with confidence from the mixing ratio of H_2 to heavy gas.

H_2 Blowoff

The instability of an H_2 atmosphere follows from its large scale height $H_1 = kT/m_1 g$, which is not much smaller than the radius r_o of Titan. A convenient variable is λ, which is essentially the ratio of gravitational and thermal energies:

$$\lambda = \frac{GMm_1}{kTr} = \frac{r}{H_1} \qquad (1)$$

The symbols are: k, Boltzmann's constant; T, temperature; m_1, the mass of an H_2 molecule; g, the acceleration of gravity; G, the gravitational constant; M, the mass of Titan; r, the distance from its center. My illustrations were for a somewhat arbitrary constant temperature, T = 100°K, but are not grossly changed for other likely temperatures. For this temperature, $\lambda_o = 8.6$ at the surface, and it varies as $1/r$. Indeed, the hydrostatic equation for the number density n can be written

$$n = n_o e^{\lambda - \lambda_o} \qquad (2)$$

Even at an infinite distance $\lambda_o - \lambda$ is only 8.6, and the density is only reduced by the factor exp(8.6) = 5400. The static situation is unstable, and will be replaced by a planetary wind, whose description resembles that of the solar wind. In that phenomenon, the speed is found to be sonic at $\lambda = 2$, a distance of 4.3 r_0 or 11,000 km from the center of Titan. The time constant for this blowoff is found to be only 4 hours.

This description is unlikely to be modified by changes in the temperature profile, unless the expansion is so rapid that adiabatic cooling overcomes conduction and solar heating to produce boundary temperature of 10-20°K...

Danielson: It will fall back in then.

Hunten: Yes, perhaps as clouds of condensed hydrogen (Sagan 1973). I don't think the situation is stable, because without very rapid outflow the adiabatic cooling cannot overcome the heat sources. A preprint by Gross (1973) begins to attack this problem by modeling the atmosphere as a polytrope. In a sense, my adoption of a constant temperature implies an exact balance of adiabatic cooling with solar heating.

Pollack: The normal balance in a planetary atmosphere is between heating by the Sun and radiation in the infrared; I really don't understand this adiabatic atmosphere.

Hunten: That's normally true; but I suspect that adiabatic cooling would become important as well in this blowoff situation. I don't really believe in the pure hydrogen atmosphere anyway, because it would require a huge source.

Strobel: Adiabatic cooling can be important if the time scale of the expansion is short compared to the time constant for heating.

Hunten: That's the idea; the time scale for this flow is less than an Earth day. If the same reasoning is applied to heavier and heavier gases, with their larger values of λ, stability against blowoff is found for masses greater than 6, although evaporation by the Jeans process can still take place. Thus, methane, with its mass of 16, is safe unless the temperature is considerably greater than 100°K. This temperature is a kind of mean for the whole atmosphere, to be used in the barometric equation.

Trafton: The barometric equation doesn't apply in such a non-equilibrium situation.

Hunten: That's true; properly speaking, the acceleration of the gas must be included, and we get the equation of motion used for the solar wind. But the variation of density with height isn't much different, especially for the heavier gases that are stable, or almost stable, against blowoff.

111

Now, to turn to the case of a mixture of H$_2$ with a heavier gas: if the composition doesn't vary with height, the mean mass can be used to test for blowoff. Thus, a 50:50 mixture (by number) of H$_2$ and CH$_4$, with a mean mass of 9, should be stable. But what about the tendency of the H$_2$ to rise through the CH$_4$, take up its own scale height, and approach the state of diffusive equilibrium? This is the really new idea in Paper I, where I introduce the idea of "limiting flux". (Actually, the idea has been used for terrestrial hydrogen for years; the new thing is just the application to other situations.) The H$_2$ escapes freely from the top of the atmosphere; thus, it is like a gas flowing through a barrier into a vacuum. We must assume an equal source in the lower atmosphere or in the interior of Titan. The resistance of the methane (or other gas) to flow is inversely proportional to the diffusion coefficient $D_1 = b_1 (n_1 + n_a)$. Here b_1 is a constant, the "binary collision coefficient", and n_1 and n_a are the number of densities of H$_2$ and the background atmosphere. The resistance therefore is less at greater altitudes, and the same H$_2$ flux is carried by a smaller density at a higher speed. In the limit of easy escape from the top, the result is a "limiting flux" ϕ_ℓ and a constant mixing ratio $f_1 = n_1/n_a$. (Note that f_1 is the ratio of H$_2$ to everything else, not to the total.) Specifically,

$$\phi_\ell = \frac{b_1}{H_a} \cdot \frac{f_1}{1+f_1} \left(1 - \frac{m_1}{m_a} \right) \tag{3}$$

Unless f_1 approaches or exceeds unity, this flux is proportional to f_1, and therefore really represents a "limiting velocity", inversely proportional to n_a as just discussed.

In Paper III, Strobel and I relate this simple, but general, description to a detailed model for the Earth's atmosphere, and find that the description is very satisfactory. If there is a temperature gradient, Equation 3 includes another small term involving thermal diffusion (Paper II).

If the actual flux is less than ϕ_ℓ, the H$_2$ density does not fall off as rapidly with increasing height, and much larger densities are obtained at great altitudes. This is the way the system tends to react if escape from the top is less than easy; the densities are larger and tend to maintain almost the same flux. The Jeans equation tells us the density required for a given flux, not the flux that is obtained for a given density. If escape is still more difficult, the system reverts to the usual description, in which the flux is derived from the density. But this situation is highly improbable for H$_2$ (and He) on Titan.

If the limiting flux is strictly realized, so that the mixing ratio is independent of height, then it does not matter whether or not the atmosphere is vertically mixed by eddy diffusion or other forms of vertical motion. These processes tend to produce a constant mixing ratio too, but they make no difference if that is already the case. Usually we lump all the mixing processes into an eddy coefficient, whose value is always a problem when we study an upper atmosphere. For our purpose, it is irrelevant.

Danielson: Surely that's only true up to some limiting mixing rate?

Hunten: Yes - if the mixing is too intense, there may be no part of the atmosphere where Equation 3 can be applied to get the actual flux. This is true for helium, and heavier gases, on Earth. I don't want to go into details now, but there is a discussion in Paper II, with a criterion to tell which situation applies to a given gas.

Pollack: In the lower part of the atmosphere, where mixing dominates over diffusion, how do you get a value for the flux?

Hunten: You're perfectly right -- in such a region the flux can be almost anything, but the H_2 mixing ratio is constant. To find the flux, we must find a region dominated by molecular diffusion, namely the region above the homopause (or turbopause). Some people call it the heterosphere, others the diffusosphere. In this region Equation 3 can be applied to get the flux. Paper I gives values of the coefficient that multiplies $f_1/(f_1 + 1)$; it is a little under 2×10^{12} cm^{-2} sec^{-1} for a CH_4 atmosphere, and twice as great for N_2. The temperature dependence is very slight, because b_1 varies as $T^{0.8}$ and H_a as T. If we focus on the methane atmosphere, we obtain the following typical limiting fluxes, in molecules cm^{-2} sec^{-1}:

f_1	0.1	1.0	2.5	10
ϕ_ℓ	1.7×10^{11}	9×10^{11}	1.3×10^{12}	1.7×10^{12}

The third case uses the 5 km-A of H_2 from Trafton (1972a) and the 2 km-A of CH_4 from Trafton (1972b). The second is the composition favored in Pollack's (1973) greenhouse model. The last entry, with a mean mass of 3.3, would have to be examined critically, because it is subject to blowoff. Even though the H_2 is blowing away, it does not necessarily carry along the CH_4, because they can part company at very high altitudes (Paper II).

Paper I contains a model of the outer atmosphere, or corona, for the modest flux of 1.3×10^{11} cm^{-2} sec^{-1}. The methane is confined to low altitudes, and above that the H_2 takes up its own scale height. At 4.3 Titan radii, where $\lambda = 2$, is the critical level, and the effusion velocity is 100 m/sec. Thus, the loss is still slightly short of being hydrodynamic. If the corona is not isothermal, as we discussed earlier, the critical level will adjust itself, but I don't expect any major difference.

Possible Sources of H_2

I am convinced that the fluxes shown in the table above are correct for the mixing ratios assumed. If we believe Trafton's data or Pollack's model, we must also believe that Titan has a source of H_2 amounting to 10^{12} molecules cm^{-2} sec^{-1}. If we do not like that, we must add something like the 50 km-A of N_2 that I suggested last year (Hunten 1972). The flux is then 1.7×10^{11} cm^{-2} sec^{-1}, which is still large.

Danielson: You can live with Trafton's abundances?

Hunten: Yes, if the flux is acceptable.

Trafton: But those abundances are very uncertain. There could be less methane if some other gas is present, and the laboratory calibration is another source of error. For H_2, the error could easily be a factor of 2 because it is hard to determine the continuum position.

Hunten: Anyway, your data and Pollack's greenhouse model seem to tell us to look for a strong source of H_2. Methane photolysis begins at 1550 Å, and the Sun provides only 10^{10} photons cm^{-2} sec^{-1} at shorter wavelengths, too small by a factor of 100...

Sagan: What about indirect photolysis, sensitized by some other molecule?

Hunten: Perhaps, but we can also think about NH_3. Paper I contains some estimates, based on a suggestion by Lewis (1971). There should be a small amount of ammonia near the surface, and it could just provide the source if everything worked at top efficiency. The solar radiation below 2250 Å would have to reach the surface, and the quantum yield for H_2 production would have to be high. Neither seems really probable to me, and Lewis has been telling me the same thing.

I am much more disposed to favor a source in the interior of Titan. Lewis' models suggest the presence of a strong NH_3-H_2O solution, which is fairly corrosive and might react with other things to make H_2. Again, it seems improbable but perhaps we should not rule it out just yet. Radiolysis by α, β, and γ emissions from radioactive decay is also a possibility. While a flux of 10^{12} cm^{-2} sec^{-1} seems large for an atmosphere, it is probably trivial for an interior.

I notice Lewis' eyebrows getting higher and higher, and he may want to say something. Nevertheless, I think we are logically forced to believe in the presence of some potent H_2 source, if we believe that large amounts of it are present in the atmosphere.

Possible Recycling from the Toroid around Saturn

What about the influence of the toroid of gas orbiting around Saturn, suggested by McDonough and Brice (1973) and the preprint by Sullivan (1973) that just reached us? Both papers point out that hydrogen can be recycled back to Titan's atmosphere, and suggest that the net loss rate may be much less drastic than suggested here and by others. I have not considered the question adequately, but I suspect that the response of Titan to such recycling will be to increase its outward flux to compensate, until the difference between the outward and recycled fluxes is again equal to ϕ_ℓ. The mechanism is to build up the coronal density by whatever factor is needed to raise the outward flux. If the net flux is much less than ϕ_ℓ, the scale height of H_2 above the homopause will be large, instead of being equal to that of CH_4. Efficient recycling would probably require the density of the toroid to approach the density at infinity required by (2), with the reference density n_0 taken at the homopause. Toroid densities of 10^{10} cm^{-3} or more might be required, far greater than any thing suggested by McDonough and Brice. Sullivan, on the other hand, adopts an approach similar to the one just outlined, and obtains a density as high as 10^{11} cm^{-3}.

McDonough: What size atmosphere are you talking about?

Hunten: In my example, the diameter is 8.6 Titan radii, big enough to sweep up quite a lot of hydrogen in your model. But the problem I see is that your toroid may need a very high density to be effective in recycling. Your model does not include the collisions between molecules, and they will dominate the situation if the density is as high as I suspect. Perhaps the collisions can be included, but the final model may be a very different one.

Sagan: A remark on the source of the hydrogen. Experiments in prebiological organic chemistry start with methane, ammonia, and water, but no hydrogen. As the irradiation proceeds, and organic molecules build up, and a substantial fraction of the initial mass of hydrogen becomes free H_2. If we believe that organics are made this way on Titan, we must also believe that H_2 is being produced.

Hunten: Certainly. But there is still a question about the total rate of production. At unit quantum efficiency, the process must use all the solar photons up to 2300 Å to get the inferred fluxes of H_2.

Sagan: But you can also do it by inefficient photolysis up to 3400 Å, as in our experiments, and that's not out of the question. The other possibility is to let the solar radiation drive the weather and produce thunderstorms. Visible light can do this and let you draw on a much larger energy source, converted inefficiently into electrical discharges.

Now, what about the material left behind by the escaping hydrogen? If 10^{11} to 10^{12} molecules cm^{-2} sec^{-1} escape from Titan over geological time, how much ice is volatilized? By a quick calculation, the column seems to be several kilometers deep...

Hunten: Or a comparable layer of organic compounds (see Paper I).

Sagan: Now, the idea of melting a few kilometers or a few tens of kilometers of planetary surface over geological time is not unknown to us on the planet Earth. I don't find anything bizarre about a similar turnover on Titan. What about methane and ammonia volcanoes, with liquid ammonia lava, to help with outgassing?

Lewis, Sagan: (Discussion of such volcanoes and the possibility of observing them through breaks in the clouds. Lewis feels that volcanic activity is at best no easier to generate under Titanian conditions than terrestrial ones.)

Hunten: At any rate, volcanoes will merely release gas into the atmosphere. They don't help make hydrogen out of it.

Lewis: I just did a quick calculation about radiolysis in the interior. If all the energy from radioactive decay in Titan were deposited in the right kind of material, and with the usual rule of thumb that 1 percent of it goes into new chemical bonds, there is no problem in making enough hydrogen for you. In fact, there is a factor of 1000 to spare.

Hunten: Provided the radioactive material is not segregated from the H_2O, NH_3, and CH_4.

Lewis: There should be plenty of water of hydration right in the silicates, provided the internal temperature did not get high enough to dehydrate silicates and physically separate all the water.

(Post-meeting Note: Regrettably, a later check on this estimate showed it to be far too optimistic. Lewis takes a chondritic radioactive energy release of 5×10^{-8} erg gm^{-1} sec^{-1} diluted by 3 times the mass of ices. If 25% of the energy is in 1 MeV β particles, and if they have a 1% radiolysis efficiency, the production rate per cm^2 of surface is 5×10^8 H_2 molecules. While this estimate might be raised somewhat by a detailed calculation, it is far short of the desired 10^{11} - 10^{12} cm^{-2} sec. The N_2 source mentioned below by Lewis is still of interest: the flux estimated above, divided by 3, would give an accumulation of 10 km-atm.)

Trafton: What about other icy satellites; shouldn't hydrogen be produced in them by the same process?

Hunten, others: Some satellites may not have internal convection to bring the gas near the surface. In any case, unless there is an abundant heavy gas, like methane, to retard the loss of H_2, a measurable density of hydrogen cannot be built up.

Lewis: According to experiments on, for example, organic crystals, radiolysis by 1 MeV β particles is almost 100 percent efficient in initially breaking chemical bonds. Many of them then recombine, but new compounds are also formed.

One last word on radiolysis. If it is taking place, I expect water to be the compound that is broken up to make H_2. The oxygen will react with ammonia rather efficiently to make nitrogen.

Danielson: Then why don't we get the same thing with the Galilean satellites? We should see nitrogen there.

Pollack: Yes, why don't some of the Jupiter satellites have nitrogen in their atmospheres? For two of them, there are occultation experiments that have set very severe limits on the surface pressure.

Lewis: There are several possible answers; a satellite may not contain ammonia, or the radioactivity may be segregated from the water and ammonia. Several conditions must be satisfied for radiolysis to be effective. I am only giving a plausibility argument for a source of H_2 and N_2.

Blamont: But several satellites may contain a hydrogen source?

Lewis: Yes, if only by radiolysis of water.

2.10 GASEOUS TOROID AROUND SATURN

Thomas R. McDonough

Introduction

It has been suggested (McDonough and Brice 1973a) that Titan's escaping atmosphere could be trapped in the Saturnian System in the form of a toroidal ring or "doughnut". The radius of the toroid would be comparable to Titan's orbit, or about ten times larger than the visible rings. Theoretical analyses of the toroid have been made by McDonough and Brice (1973b), Dennefeld (1973), Tabarié (1973), and Sullivan (1973).

The condition that the majority of escaping atoms or molecules from Titan be unable to escape Saturn's gravitational attraction is that the exospheric thermal velocity be less than or comparable to the orbital speed of Titan, V_T, or numerically,

$$T_{ex} \lesssim 1200m \tag{1}$$

where m is the atomic or molecular weight, and T_{ex} is the exospheric temperature in °K. Thus, if the exospheric temperature is within an order of magnitude of Trafton's (1972a) Titanian exospheric temperature of 74°K, most of the escaping atmosphere will orbit Saturn regardless of its composition.

The possibility of much higher exospheric temperatures has been suggested independently by Gross and Mumma (1973) in which case the temperature might violate Equation 1. Then, while most of the gas might escape Saturn, some fraction of it would still be retained, with the Titanian escape flux still yielding a substantial amount of gas orbiting Saturn. However, the models presented here at the Workshop predict temperatures that easily satisfy Equation 1.

A constraint on the composition and density of a Saturnian toroid is given by the paper of Franklin and Cook (1969). They studied principally the constraint on the amount of sodium vapor surrounding Saturn's visible rings. They derived an upper limit of $\sim$100 Na atoms cm^{-3} for a gas ring comparable in size to the visible rings, which yields an upper limit $\sim$10 times smaller for the proposed toroid. Their paper placed no constraints on the probable constituents of the toroid (H, H_2, CH_4, etc.).

Size of the Toroid

Characteristic inner and outer radii of the toroid, $r_\pm$, are given, from celestial mechanics, by:

$$\frac{r_\pm}{r_T} = \left[2\left(\frac{V_T}{V_T \pm V_{th}}\right)^2 - 1 \right]^{-1} \tag{2}$$

where r_T is Titan's orbital radius and V_{th} is the RMS thermal speed of the particle immediately after escape from Titan. The thickness of the toroid, ΔZ, perpendicular to the equatorial plane, is:

$$\Delta Z/r_T \simeq 2\ V_{th}/V_T \tag{3}$$

For 70°K hydrogen atoms (or 100°K hydrogen molecules or 280°K methane molecules), r_-/r_T, r_+/r_T, and $\Delta Z/r_T$ are 0.4, 3, and 0.5, respectively. A schematic top view of this toroid is shown in Figure 2-38.

Hydrogen Density in the Toroid

If the average toroid particle survives for a period of time T, and if we estimate the volume of the toroid to be $\sim r_T^3$, then its mean density, N_T, is:

$$N_T \sim FT/r_T^3 \tag{4}$$

where F is the atmospheric escape rate from Titan. Trafton (1972a) estimates F to be from 2×10^{26} to 4×10^{29} H_2 molecules sec^{-1} and Hunten's (1973a) maximum escape flux is slightly smaller than the latter figure. If we take as a characteristic lifetime the ionization lifetime of hydrogen atoms at Saturn's distance, T is 6 yr and Equation 4 implies densities from ~ 10 to $\sim 10^4$ molecules cm^{-3}.

Tabarié (1973) independently estimates, for a series of Titanian model atmospheres, escape fluxes of hydrogen molecules from 5×10^{27} to 9×10^{32} mols sec^{-1}. The associated atomic hydrogen escape fluxes range from 8×10^{26} to 10^{27} atoms sec^{-1}. It would appear from these estimates that the density of atomic hydrogen in the toroid is greater than 10 atoms cm^{-3} and would thus be optically thick at Lyman-α wavelengths. However, the above density estimates may need to be lowered because of recapture of the hydrogen by Titan as discussed below.

Recycling of Titan's Atmosphere

Particles which escape Titan but not Saturn will orbit in elliptical trajectories that intersect Titan's orbit, unless they are perturbed. Since the particle can orbit Saturn many times, Titan and the particle may happen to pass through the same region at the same time, allowing the particle to be recaptured. The effective area of Titan for recapture of toroid particles may be much larger than Titan's visible area because its exospheric radius may be an order of magnitude larger than the visible radius (Trafton 1972a, Hunten 1973a), due to Titan's weak gravitational field. It is estimated (McDonough and Brice 1973a,b), using a simplified orbital model, that up to $\sim 98\%$ of 100°K hydrogen molecules could be recaptured in this way, if the radius of the base of Titan's exosphere is of the order 2.5×10^4 km. It remains to be investigated whether interparticle collisions, gravitational perturbations, or radiation pressure significantly alter the orbits to hinder this atmospheric recycling phenomenon.

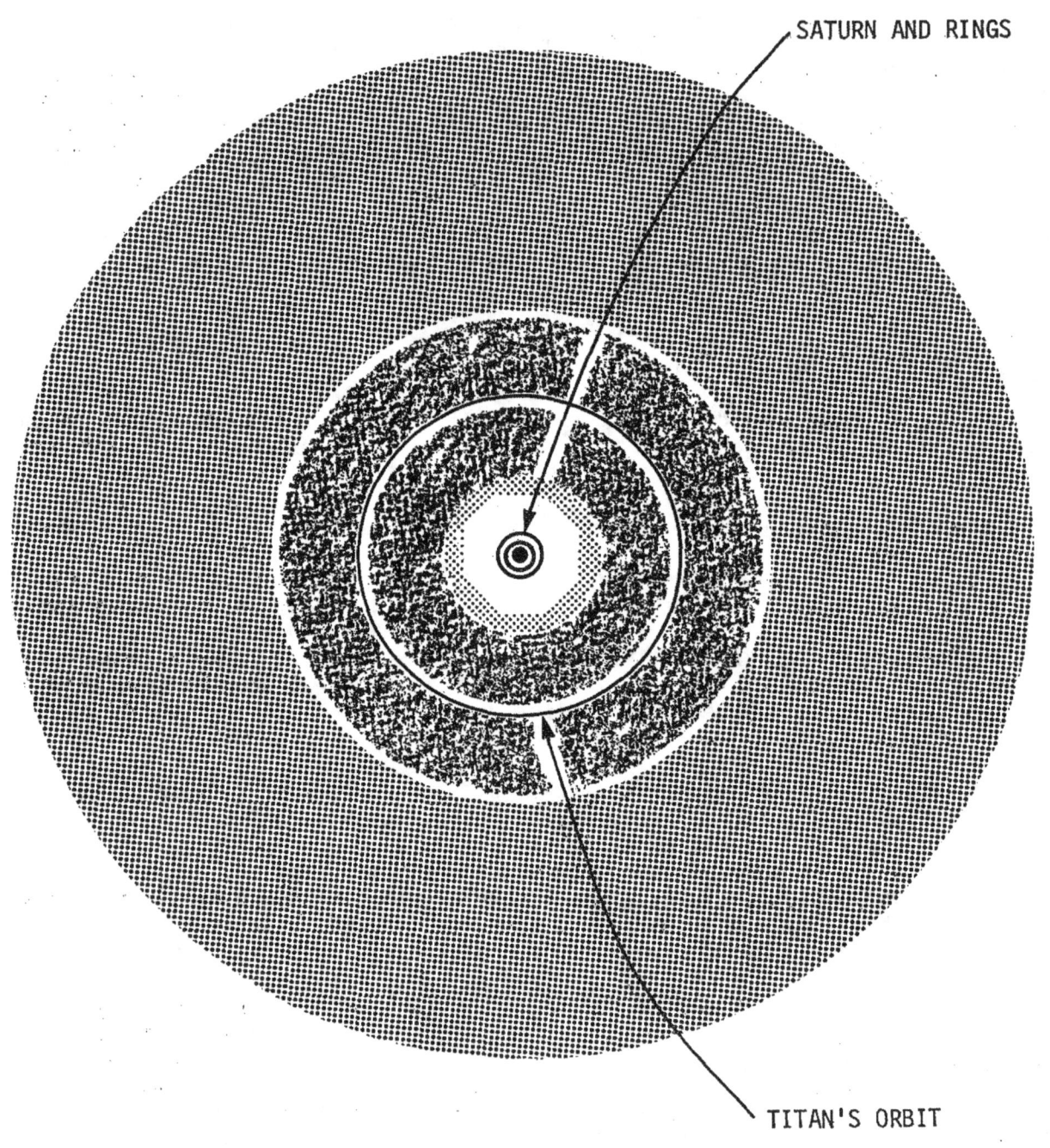

SATURN AND RINGS

TITAN'S ORBIT

├──────────┤ SOLAR DIAMETER

Figure 2-38.   The Titan toroid, as viewed from above, is represented by the
shaded areas. The boundaries shown are approximate, representing regions of different densities. These boundaries are estimated for 100°K H_2, and are diffuse. Saturn, the visible Rings,
Titan's orbit, and the Sun's absolute diameter are shown
approximately to scale. After McDonough and Brice (1973).
Reprinted from Icarus, 20:138, with permission of Academic
Press, Inc. Copyright © 1973 by Academic Press, Inc. All
rights of reproduction in any form reserved. Printed in
Great Britain.   120

Effects of a Saturnian Magnetosphere

If Saturn has a magnetosphere similar to Jupiter's, which extends to the vicinity of Titan's orbit, it will protect toroid particles from direct ionization by solar-wind particles, although some ionization by solar-wind particles that penetrate the magnetopause and diffuse to the toroid is possible. Photoionization and photodissociation would probably be important. Furthermore, corotating magnetospheric plasma could ionize the toroid particles. Observation of the toroid density as a function of radius could thus provide information on the presence or absence of a Saturnian magnetosphere.

Scientific Usefulness of the Toroid

The composition of the toroid would reflect the composition of Titan's atmosphere, and could provide information (possibly even from Earth orbit) on whether the Titanian atmosphere is in a condition of blow off, on whether hydrogen is present in substantial amounts in the atmosphere of Titan, and on the ratio of atomic to molecular hydrogen. If the toroid can be mapped, e.g., by ultraviolet photometry, the Titanian exospheric temperature and net loss rates may be found from Equations 2 through 4. The density of the toroid as a function of radius and azimuth could provide information on day-night asymmetries at Titan, the possible Saturnian magnetosphere, and the rotation rate of Titan.

This article is, in part, a summary of publications by McDonough and Brice (1973a,b). The research was sponsored, in part, by the NASA Physics and Astronomy program under Grant no. 33-010-161, and the National Science Foundation Atmospheric Sciences Section under Grant nos. GA-11415 and GA-36916.

Pollack: Can we compare Trafton's and Hunten's escape fluxes? Were they referred to the same level?

Trafton: Mine were shown for both the critical level and the surface; the total escape rate was up to 5×10^{29} molecules/sec.

Hunten: And mine were all referred back to the surface. If we take a typical flux of 10^{11} cm^{-2} sec^{-1}, the total rate is almost 10^{29} sec^{-1}, so we are basically in agreement. However, I remind you that my calculation depends only on the rate of diffusion. If the exosphere temperature is changed, or hydrogen is recycled from the toroid, the structure of the corona changes to maintain the same net loss rate.

Strobel: What is the time required to fill the toroid to equilibrium, if it started out empty?

McDonough: The average atomic-hydrogen lifetime in the toroid is about 150 orbits or 6 years and is determined by the rate of ionization of the hydrogen. The time to fill the toroid would be of this same order, unless recapture is important.

Hunten: Regarding your estimate of up to 98% of the hydrogen being recaptured; if that were to occur, the lower net escape flux will increase the exospheric density, hence, re-establishing the original net escape flux. The flows may be more hydrodynamic than molecular at this point, which could further inhibit your suggested high recapture flux.

Regarding perturbations of the atomic orbits, I calculated some time ago that for a particle density in the toroid of 10^3 particles cm^{-3}, the mean free path is about equal to the toroidal circumference. I am not sure what this mean free path does to the density distribution but it should randomize the orbits rapidly.

Sagan: Would the toroid have a detectable effect on the solar wind?

McDonough: Yes, it has a depth of one mean free path for charge exchange when the density is 10^3 cm^{-3}. So it will cast a shadow if it is at least that dense.

Pollack: How bright would the toroid be in Lyman-α?

Blamont: Maybe as much as 500 Rayleighs which is somewhat greater than the 300 Rayleigh diffuse background glow.

McDonough: I think the scattering from the toroid could be distinguished if there were enough spectral resolution to separate the components by their Doppler shifts. An OAO-type spacecraft would be ideal for looking at the toroid, although the field of view and the slits are probably too narrow on the Copernicus spacecraft.

Caldwell: The first OAO did try to detect Lyman-α from Saturn, but as I recall, it did not succeed for several reasons.

2.11 ATOMIC HYDROGEN DISTRIBUTION

N. Tabarié

Introduction

Molecular hydrogen (H$_2$) has been identified in Titan's atmosphere by its quadrupole absorption line, with an estimated total quantity of 5 km-A (Trafton 1972). There must then be some atomic hydrogen (H) in Titan's atmosphere, which may be accessible to observation in Lyman-α at 1216 Å.

In this study we have examined several possible H$_2$ vertical distributions with the constraint of 5 km-A as a total quantity, and have calculated, with approximations, the corresponding vertical distributions of atomic hydrogen. It was found that the H distribution is quite sensitive to two other parameters of Titan's atmosphere: the temperature and the presence of other constituents. The escape fluxes of H and H$_2$ were also estimated as well as the consequent distributions trapped in the Saturnian system.

Models of H$_2$ Distribution

The constraint of 5 km-A of H$_2$ makes impossible an atmosphere of pure H$_2$, since it would escape so fast that the total mass of Titan would vanish in less than 10^6 years. Additional constituents with molecular weight higher than H$_2$ have to be present to accommodate the 5 km-A of H$_2$. Two constituents were considered separately: (1) N$_2$ after a suggestion by Hunten (1972), with various mixing ratios ranging from 10^{-3} to 10^{-1} for H$_2$/N$_2$; and (2) CH$_4$, which was positively identified by Kuiper (1944), with a mixing ratio of 1.

For the sake of simplicity the atmosphere was considered to be spherically symmetrical and isothermal, with a temperature of either 80, 100, or 120°K. By analogy with the terrestrial atmosphere, three different zones were considered: The turbosphere where the mixing is constant, the diffusosphere where each constituent follows its own scale height, and the exosphere where collisions are not important. The turbopause level was defined as the altitude Z_a where the total number density is 10^{11} mols cm^{-3}. The exobase level was defined as the altitude Z_c where the density of H$_2$ is 10^7 mols cm^{-3}, except if it was found to be higher than the blow off level (kinetic energy equal to gravitational energy), in which case the exobase was located at the blow off level. These levels are indicated in Table 2-8, together with the escape fluxes of H$_2$ at the exobase estimated from the Jeans formula. The various calculated H$_2$ distributions are indicated in Figure 2-39. The main feature of these H$_2$ distributions is the existence of a very large diffusosphere when compared to the terrestrial case. This is due to the low gravity of Titan and to the high total quantity of H$_2$.

Models for H Vertical Distribution

Photodissociation of H$_2$ and CH$_4$ by the solar flux was considered as the only source of H, compensated by escape at the exobase and by three-body recombinations at lower altitudes. For each H$_2$ profile a corresponding H profile

Table 2-8a. H_2 Fluxes for Titan Atmospheric Models

TEMP.	MIXING RATIO	TURBO-PAUSE Z_a km	EXOBASE Z_c km	DENSITY AT GROUND (H_2) mols cm^{-3}	DENSITY AT Z_c (H_2) mols cm^{-3}	EFFUSION VELOCITY (H_2) cm sec^{-1}	ESCAPE FLUX (H_2) mols cm^{-2}sec^{-1}	TOTAL FLUX (H_2) mols sec^{-1}	FLUX AT GROUND (H_2) mols cm^{-2}sec^{-1}
80°K (H_2-N_2)	10^{-3}	520	1570	11.0×10^{18}	10^7	2.3×10^2	2.3×10^9	4.8×10^{27}	5.9×10^9
	10^{-2}	470	3420	10.5×10^{18}	10^7	1.3×10^3	1.3×10^{10}	5.9×10^{28}	7.0×10^{10}
	10^{-1}	460	8950	9.5×10^{18}	10^7	7.0×10^3	7.0×10^{10}	1.6×10^{30}	2.0×10^{12}
100°K (H_2-N_2)	10^{-3}	670	2410	8.0×10^{18}	10^7	1.5×10^3	1.5×10^{10}	4.6×10^{28}	5.6×10^{10}
	10^{-2}	610	6050	8.0×10^{18}	10^7	6.8×10^3	6.8×10^{10}	6.3×10^{29}	7.7×10^{11}
	10^{-1}	590	12060*	7.0×10^{18}	3.6×10^7	1.4×10^4	5.0×10^{11}	1.3×10^{31}	1.6×10^{13}
120°K (H_2-N_2)	10^{-3}	840	3470	6.4×10^{18}	10^7	5.5×10^3	5.5×10^{10}	2.4×10^{29}	3.0×10^{11}
	10^{-2}	750	9600*	6.3×10^{18}	1.7×10^7	1.6×10^4	2.6×10^{11}	4.8×10^{30}	5.8×10^{12}
	10^{-1}	730	9600*	5.6×10^{18}	1.5×10^8	1.6×10^4	2.3×10^{12}	4.2×10^{31}	5.1×10^{13}
80°K (H_2-CH_4)	1	1470	15710*	2.7×10^{18}	2.7×10^8	1.2×10^4	3.2×10^{12}	1.4×10^{32}	1.7×10^{14}
100°K (H_2-CH_4)	1	2070	12060*	2.1×10^{18}	2.3×10^9	1.6×10^4	3.2×10^{13}	8.6×10^{32}	10^{15}

* Exobase = Blow-off level in this case.

124

Table 2-8b. H Fluxes for Titan Atmospheric Models

TEMP.	MIXING RATIO	TURBO-PAUSE Za km	EXOBASE Zc km	DENSITY AT Zc (H) atoms cm^{-3}	ESCAPE FLUX (H) atoms $cm^{-2}sec^{-1}$	TOTAL FLUX (H) atoms sec^{-1}	FLUX RATIO H/H_2
80°K (H_2-N_2)	10^{-3}	520	1570	7.50×10^4	3.7×10^8	8.0×10^{26}	1.6×10^{-1}
	10^{-2}	470	3420	2.50×10^4	2.7×10^8	1.2×10^{27}	2.0×10^{-2}
	10^{-1}	460	8950	7.80×10^3	1.7×10^8	2.8×10^{27}	1.7×10^{-3}
100°K (H_2-N_2)	10^{-3}	670	2410	2.30×10^4	2.9×10^8	9.0×10^{26}	2.0×10^{-2}
	10^{-2}	610	6050	7.40×10^3	1.7×10^8	1.5×10^{27}	2.0×10^{-3}
	10^{-1}	590	12060*	10^4	3.0×10^8	8.0×10^{27}	6.0×10^{-4}
120°K (H_2-N_2)	10^{-3}	840	3470	1.10×10^4	2.5×10^8	1.1×10^{27}	6.5×10^{-3}
	10^{-2}	750	9600*	5.60×10^3	1.8×10^8	3.3×10^{27}	6.8×10^{-4}
	10^{-1}	730	9600*	1.60×10^4	5.5×10^8	1.0×10^{28}	2.0×10^{-4}
80°K (H_2-CH_4)	1	1470	15710*	3.00×10^4	8.0×10^8	3.3×10^{28}	2.0×10^{-4}
100°K (H_2-CH_4)	1	2070	12060*	1.25×10^3	3.7×10^7	10^{27}	10^{-6}

* Exobase = Blow-off level in this case.

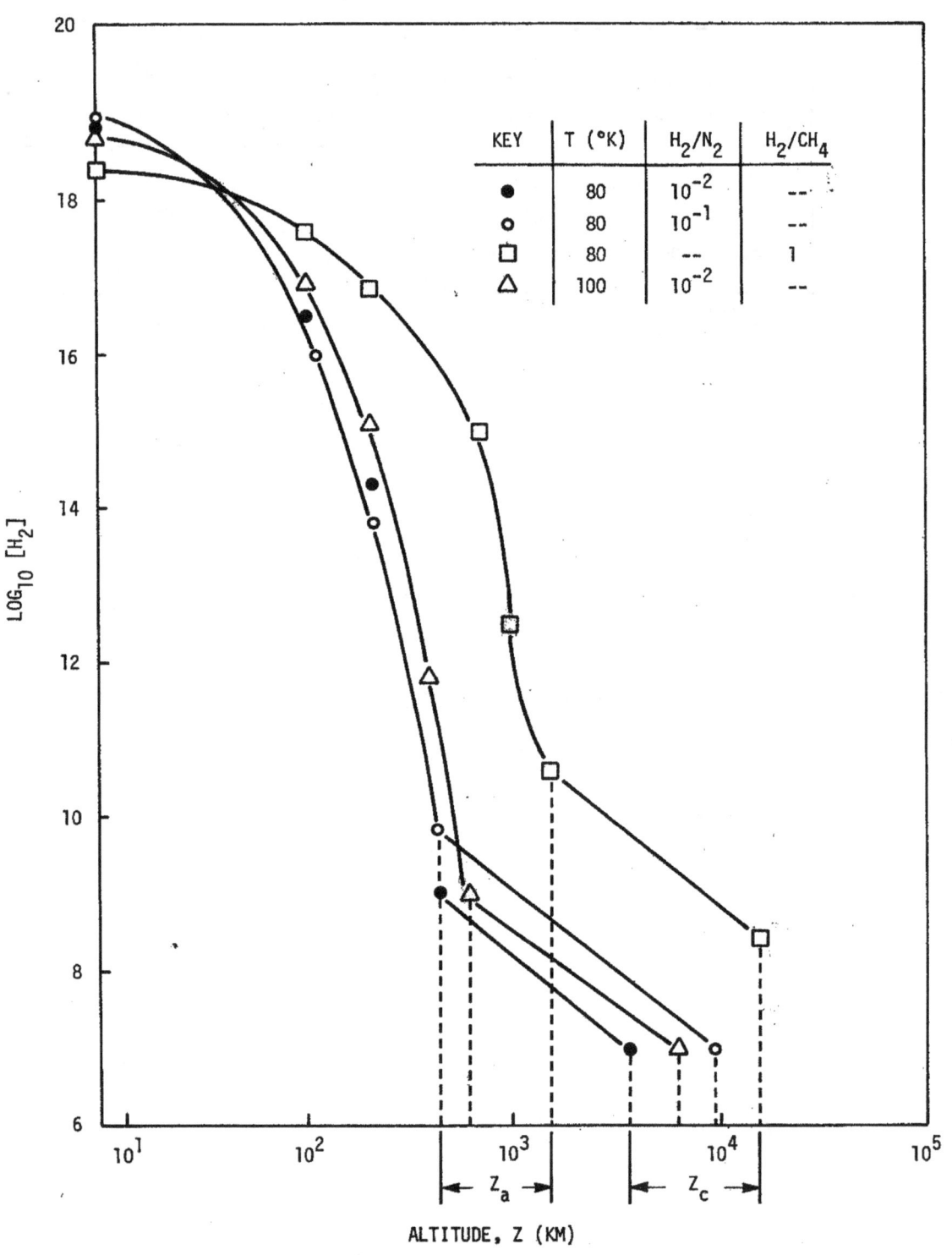

Figure 2-39.   Profile of molecular hydrogen for H_2-N_2 and H_2-CH_4 model atmospheres at temperatures of 80°K and 100°K. Z_a is the turbopause level and Z_c is the exobase level.

was calculated along the vertical at the sub-solar point by solving simultaneously the diffusion equation and the continuity equation in the spherical case:

$$\phi_H(r) = -D(r) \left[\frac{dn_H}{dr} + \frac{n_H(r)}{H(r)}\right]$$

$$\frac{d\phi_H(r)}{dr} = P(r) - L(r) - \frac{2}{r}\phi_H(r)$$

where:

$H(r)$ = scale height of atomic hydrogen.

$H_e(r)$ = scale height of the atmosphere.

$\phi_H(r)$ = the diffusion flux of atomic hydrogen in ats cm^{-2} sec^{-1},

$P(r)$ = production rate of H through photodissociation of H_2,

$L(r)$ = loss rate of H by three-body recombinations with H_2 and N_2,

$D(r)$ = diffusive coefficient:

$$D(r) = \frac{2 \times 10^{19}}{n_{H_2}(r)} \text{ in the case of diffusion of H in an}$$

atmosphere of H_2,

$$D'(r) = \frac{3.19 \times 10^{16} H_e(r)}{n_{N_2}(r)} \text{ in the case of diffusion of H,}$$

in an atmosphere essentially of N_2.

The exospheric distribution was calculated with no satellite particles from Chamberlain's theory.

For H_2-N_2 model atmospheres the effect of the mixing ratio (H_2/N_2) at ground level is illustrated in Figure 2-40 for T = 80°K. When this ratio increases the density maximum of H, located in the turbosphere, decreases, while the density of H in the diffusosphere increases substantially. Above 10^4 km altitude in the exosphere, the density is insensitive to the mixing ratio.

The effect of the isothermal temperature (T) on the H density is illustrated for an H_2-N_2 mixing ratio of 10^{-1} in Figure 2-41. The H density increases substantially with T in the diffusosphere. Again the density in the exosphere is not very sensitive to the temperature.

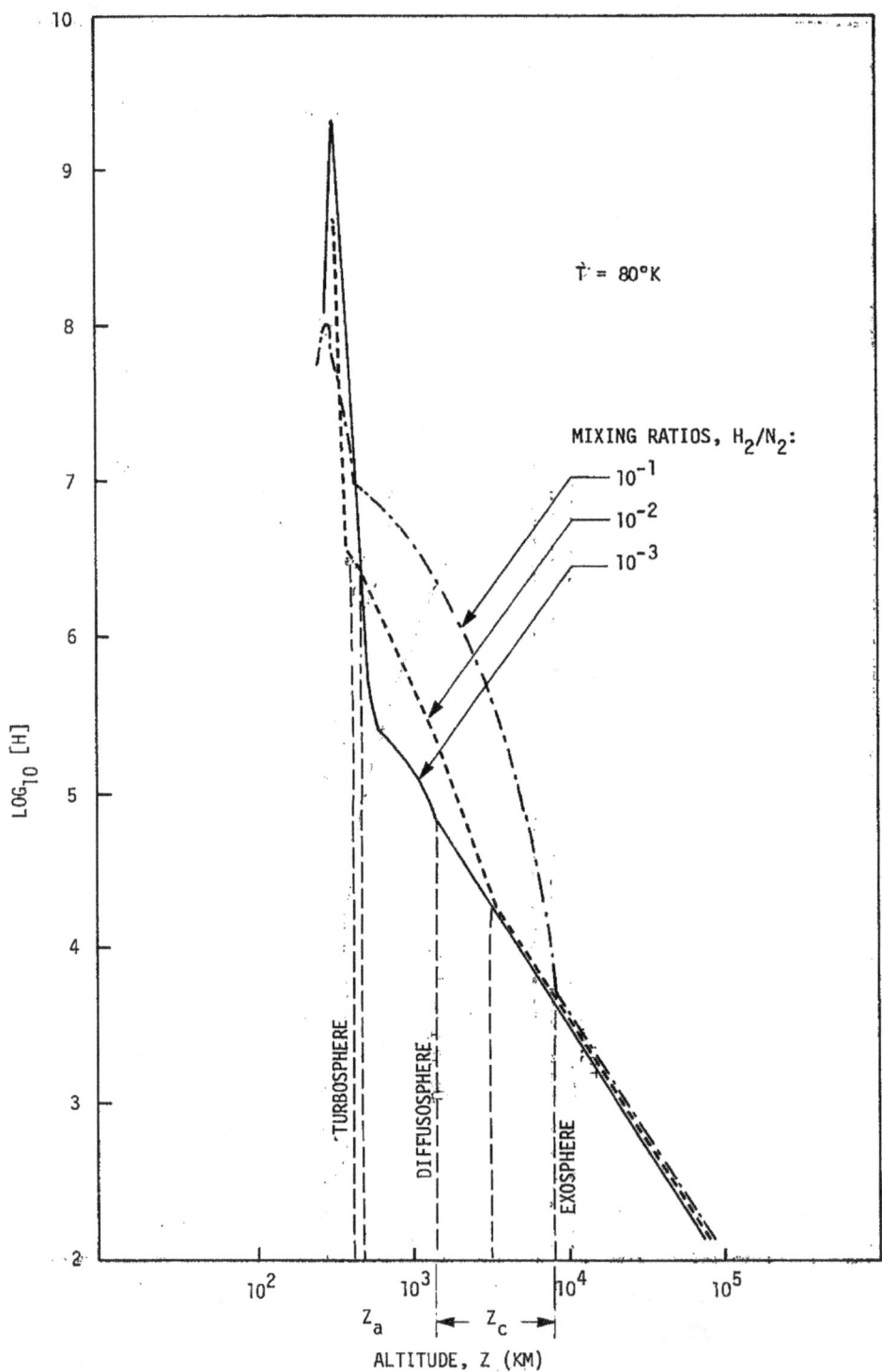

Figure 2-40. Profile of atomic hydrogen for H_2-N_2 model atmospheres at a temperature of 80°K. The effects of three different mixing ratios, H_2/N_2 equal to 10^{-3}, 10^{-2}, and 10^{-1}, are shown. Z_a is the turbopause level and Z_c is the exobase level.

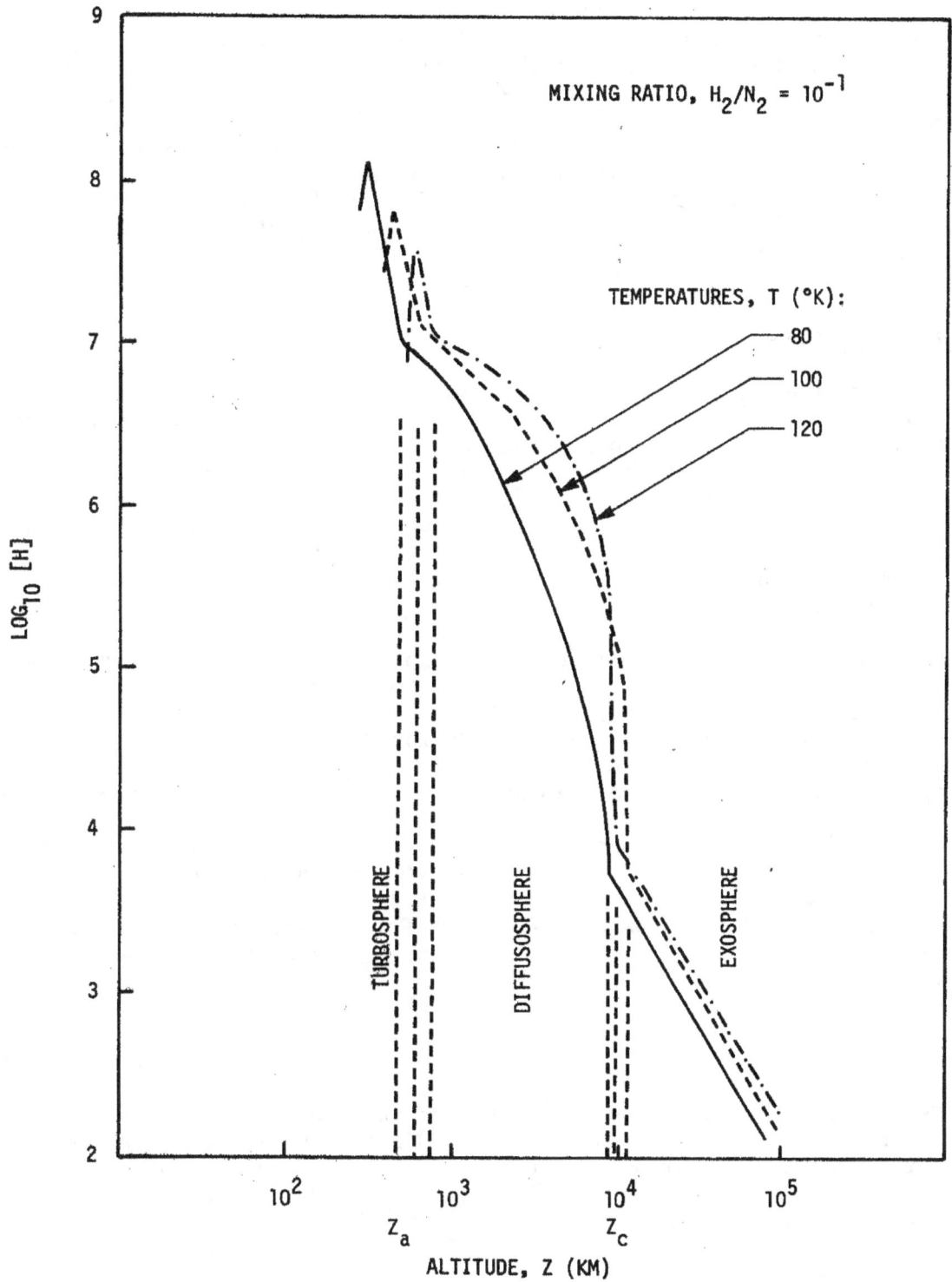

Figure 2-41. Profile of atomic hydrogen for H_2-N_2 model atmospheres for temperatures of 80°K, 100°K, and 120°K. A mixing ratio, H_2/N_2, of 10^{-1} is assumed. Z_a is the turbopause level and Z_c is the exobase level.

For H_2-CH_4 model atmospheres the effect of temperature is illustrated in Figure 2-42 for a mixing ratio of unity. The density profiles of H are quite different than for the H_2-N_2 models; the density at 5×10^3 km and on into the exosphere is much higher for the same temperature of 80°K. The effect of T is at variance with H_2-N_2 models in that an increase in T results in a decrease in the H density in an H_2-CH_4 model. This difference between the two types of models is due partly to the higher mixing ratio in the H_2-CH_4 model and partly to the increase of H production through photodissociation of CH_4. This later process is underestimated in our approach, since only photodissociation giving CH_3 + H was considered.

H and H_2 Escape Fluxes

The total fluxes of H and H_2 indicated in Table 2-8 were calculated by assuming spherical symmetry which might lead to an overestimation by a factor of 2 for H. In any case, the total flux of H is found to vary quite widely with the atmospheric parameters between 7.9×10^{26} atoms $\sec^{-1}$ for the H_2-N_2 model to 3.3×10^{28} atoms $\sec^{-1}$ for the H_2-CH_4 model at 80°K. The flux of H_2 is even larger than the flux of H for all models by a factor of between 6 and 10^6. For H_2-CH_4 models the H_2 flux is much larger than for H_2-N_2 models, and a hydrodynamic approach would be more appropriate than the assumption of hydrostatic equilibrium.

It has been suggested by Brice and McDonough (1973) that a large escape flux of H_2 from Titan would result in a toroid of hydrogen around Saturn. Dennefeld (private communication) has calculated the distribution of atomic H resulting from the escape of H and H_2. He found that a flux of 2×10^{28} atoms cm^{-3} $\sec^{-1}$ of H would result in a mean density of 22 ats cm^{-3} in the whole volume indicated in Figure 2-43 around the orbit of Titan. With a density of 22 ats cm^{-3}, such a toroid of H should be easily detected by its resonant Lyman-α emission, with an intensity of several hundred Rayleighs. In addition, since 10% of escaping H_2 molecules will be photodissociated into 2 H atoms when they are around Saturn, a flux of 10^{31} H_2 molecules (corresponding roughly to all encountered blow off conditions) would result in a much larger mean density of $22 \times 100 = 2200$ ats cm^{-3}, greatly increasing detectability of the toroid.

Summary

The distributions which we have obtained show that a determination of the vertical distribution of atomic hydrogen could easily be fitted to a model and that the two significant parameters, temperature and mixing ratio, could then be determined. However, since the differences between H_2-N_2 and H_2-CH_4 models are not very large, a high spatial resolution in the measurements would be necessary.

Hunten: Going back to my discussions of the limiting escape flux, the distributions of H and H_2 are independent of K (the eddy diffusion coefficient) until it reaches a value even larger than was assumed here (2×10^8 cm^2 $\sec^{-1}$). For very large values, the densities are decreased. (K is related to the number density at the turbopause, n_a, by $K = 2 \times 10^{19}/n_a$.) What is the storage time in the toroid for hydrogen atoms and molecules?

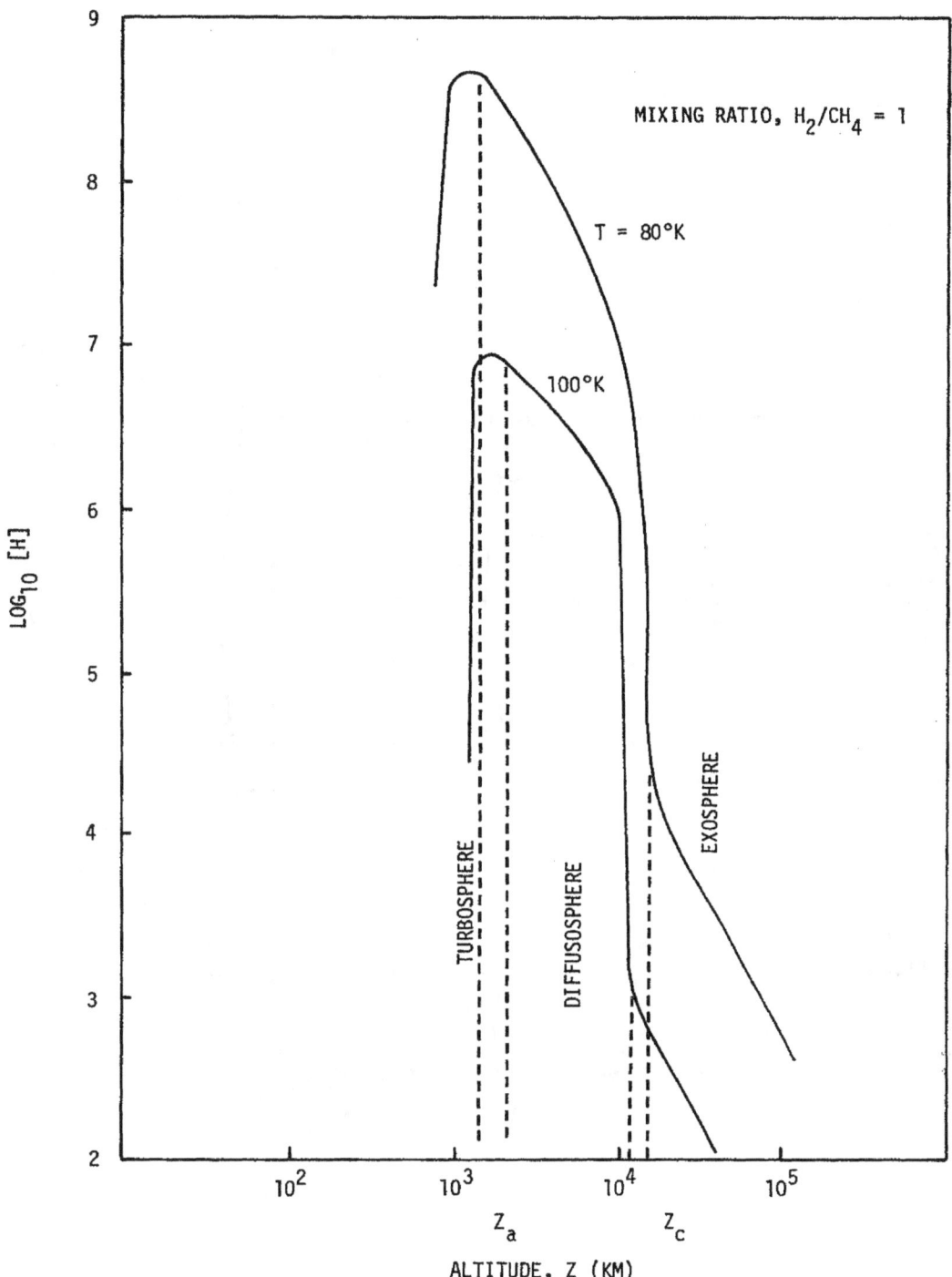

Figure 2-42. Profile of atomic hydrogen for H_2-CH_4 model atmospheres for temperatures of 80°K and 100°K. A mixing ratio, H_2/CH_4, of unity is assumed. Z_a is the turbopause level and Z_c is the exobase level.

131

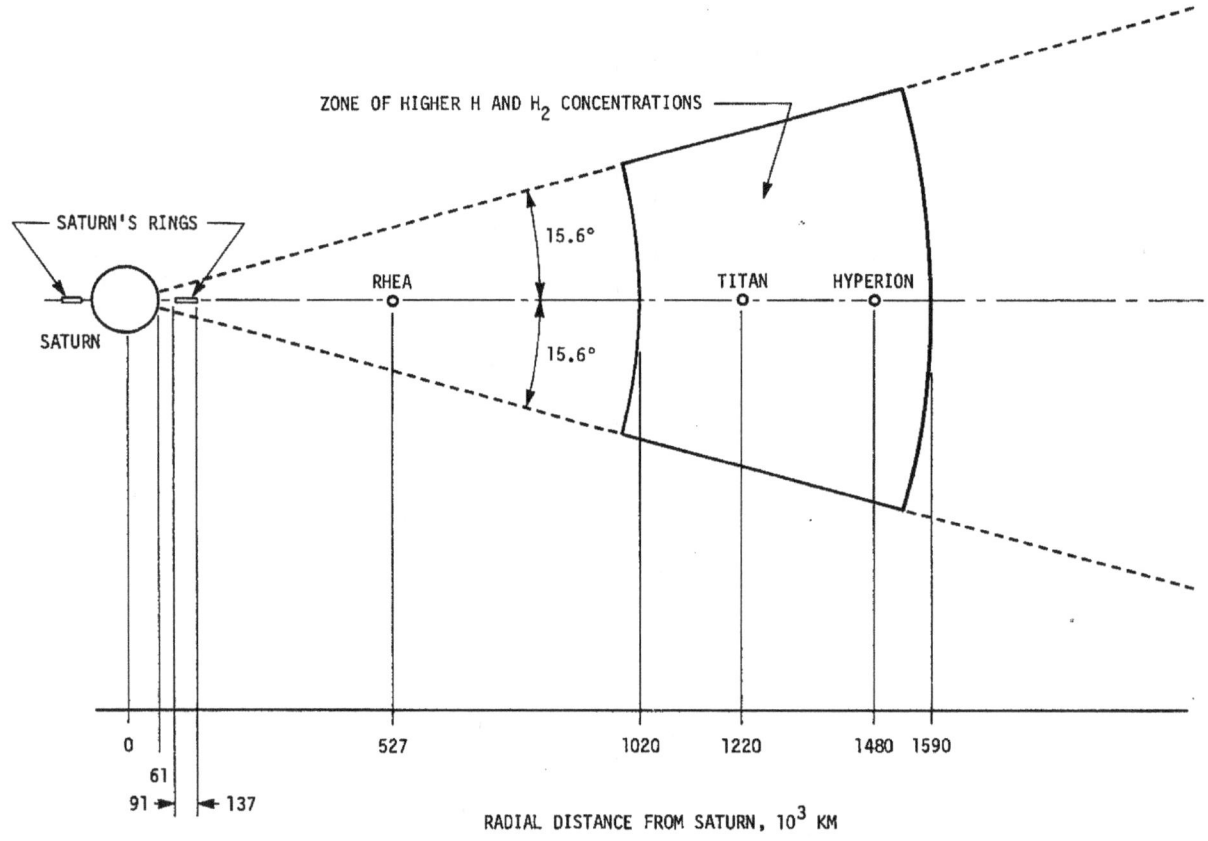

ZONE OF HIGHER H AND H_2 CONCENTRATIONS

SATURN'S RINGS

SATURN

15.6°

15.6°

RHEA

TITAN HYPERION

0

61

91 → ← 137

527

1020 1220

1480 1590

RADIAL DISTANCE FROM SATURN, 10^3 KM

Figure 2-43. Cross-section of toroidal zone H and H_2 concentrations in the Saturnian system.

Tabarié: About 2×10^8 sec which is determined by charge exchange and photo-ionization of the atoms.

McDonough: You will be able to see Lyman-α both from resonance scattering by H and photodissociation of H_2. Do you have any idea what the ratio is?

Hunten: This was an issue after the Mariner 5 flyby of Venus. Equal intensities from the two sources required an H_2/H ratio of about 10^5;

Blamont: ...which may be what we will have on Titan and in the toroid, but the line from H_2 fluorescence will be much wider and a good measurement of the line shape will allow you to discriminate.

Rasool: What will the Mariner Jupiter/Saturn (MJS) mission be able to do near Titan?

Broadfoot: MJS will have two spectrometers, one observing airglow, with a 6 arc-minute field, and one observing solar occultation (if it goes through Titan's shadow) with a 1-minute field. Occultation will give H, H_2, and CH_4. Airglow will give H, but with poorer vertical resolution.

Blamont: It seems doubtful that MJS can detect H by occultation, because the solar Lyman-α line is so much wider than the planetary absorption.

Hunten: The height resolution is not too bad if you can get close enough to Titan. One arc-minute at 30,000 km is 10 km, which isn't bad at all.

Sagan: If your instrument can measure as faint as 100 Rayleighs, it would be exciting to map the glow from the toroid or tail or whatever is there.

Broadfoot: We will be able to do that as we approach the planet; we can take days to scan over the whole Saturnian system.

2.12 ORGANIC CHEMISTRY IN THE ATMOSPHERE

C. Sagan

Introduction

There seems little doubt about the existence of an at least moderately complex organic chemistry on Titan. There is clear evidence of methane, and at least presumptive evidence of hydrogen in the atmosphere. The ratio of methane to hydrogen is the highest of any atmosphere in the solar system. The irradiation of methane/hydrogen mixtures is well-known to produce aromatic and aliphatic hydrocarbons. In addition, we know that Titan is cloud-covered and that the clouds are red. A very reasonable hypothesis is that the clouds are made of organic chemicals. I want in this paper to describe some experimental work which bears upon the possible organic chemistry of the Titanian environment.

Titan Organic Chemistry

First, suppose we have a mixture of the fully saturated hydrides of the cosmically most abundant reactive elements, hydrogen, oxygen, nitrogen and carbon. If we irradiate this mixture of methane, ammonia, water and hydrogen with short wavelength ultraviolet light -- or if we spark it in an electrical discharge, or indeed if we supply energy in any way which produces free radicals which can recombine at lower temperatures -- we produce the following simple organic compounds: the simple two carbon hydrocarbons, ethane (C_2H_6), ethylene (C_2H_4), and acetylene (C_2H_2); the simplest nitriles, hydrocyanic acid (HCN) and acetonitrile (CH_3CN); and, in the presence of water, the simplest aldehydes, formaldehyde (HCHO) and acetaldehyde (CH_3CHO). Carbon monoxide is also produced. The abundance of C_2 compounds in these experiments is variable but characteristically ranges from 10^{-2} to 10^{-6} by number, compared with the initial CH_4. These compounds are formed in experiments in which there is a great excess of hydrogen to other constituents, as well as in experiments in which the precursors are present in approximately equimolar quantities. Such experiments are characteristically performed at pressures ranging from a few tenths of a bar to 1 bar, and at temperatures from room temperature to 77°K. There does not appear to be a striking dependence of the results on the starting temperatures and pressures. In addition, all such experiments are performed in glass vessels -- but if there are wall effects, silicates are not unlikely catalysts in a planetary environment. These results are also obtained in computer quenched thermodynamic equilibrium experiments. Although the solar flux is down by a factor of 100 at Titan compared to the Earth, the ultraviolet irradiation dose may still be significant. In addition, electrical discharges in the Titanian clouds are to be expected. Therefore adequate energy sources probably exist in Titan. Also of interest is the fact that -- except for the hydrocarbons which have no permitted microwave lines -- all of the above molecules are found in the interstellar medium by radioastronomical line experiments.

Consequently there seems nothing very daring in proposing two-carbon hydrocarbons in the upper atmosphere of Titan. At lower depths, if ammonia is present, nitriles can be expected. If the surface temperatures on Titan are high enough to permit a significant vapor pressure of water vapor, aldehydes might also exist in the lower atmosphere. But these compounds are not the stable end products of the photochemistry of the Titanian atmosphere and we must inquire further about larger molecules.

Laboratory Simulation Experiments

There are of course no color photographs of Titan but the color of Titan and Saturn are not very different, and color photographs of Saturn are well-known to show a range of yellows, oranges, browns and reds. That polymeric material with such colors is readily produced under simulated Saturnian or Titanian conditions is a simple consequence of the appropriate simulation experiments. Figure 2-44 shows a characteristic experimental design from our Laboratory, in which a reaction vessel filled with precursor gases is irradiated in cylindrical geometry by the 2537 Å and in some experiments also by the 1849 Å line of mercury. The reactants are circulated Hg-free by a greaseless solenoid pump through a liquid water bath (more precisely, a NH_4OH bath) and then back again to the reaction vessel. Because hydrogen, methane, ammonia and water are all transparent at 2537 Å, these experiments use either H_2S or HCHO as the initial photon acceptor. The initial photodissociation event produces a hot hydrogen atom which is a few electron volts superthermal, and which initiates chain reactions in collision with other gases. Similar results are obtained, for example, by electrical discharges or shocks in essentially the same manner -- by producing free radicals. These particular experiments were designed to utilize the longest possible wavelengths of ultraviolet light, but any other energy source which produces free radicals should yield essentially similar results. We are at the present time engaged in experiments which more precisely simulate Titanian conditions.

Simulation Products

After a few days the reaction vessel becomes entirely coated with a reddish brown polymeric material, of which a color photograph has been published (Sagan 1971). The polymer is a flaky powder of typical particle size about 100 μm, which, on acid hydrolysis, yields a very substantial harvest of protein and non-protein amino acids. From this polymer we have isolated some 40 or 50 ninhydrin-positive peaks on the automated amino acid analyzer -- almost all of which are amino acids. We have made the first prebiological organic synthesis of the sulfur-containing amino acid cystine in such experiments. Either polynitriles are produced, which, upon solution in liquid water, give amino acids through the Strecker synthesis; or polypeptides -- polymers of amino acids -- are produced directly and hydrolyzed under acid hydrolysis.

The amino acid yields in these experiments are enormous. A good way of characterizing them is to ask what column density of amino acids would be produced after 10^9 years of ultraviolet irradiation on the primitive Earth. The answer is about 200 kg cm^{-2}, which is more than the carbon content of the sedimentary column. This simply means that amino acids are destroyed as well as made. When amino acid destruction at typical terrestrial temperatures is included, the result corresponds to about 1 kg cm^{-2} in 10^9 years, or approximately a 1 percent solution of organic compounds of amino acids in the primitive oceans. On Titan the thermal degradation of amino acids should be much slower, and probably slow enough to compensate for the factor of 100 lower ultraviolet flux.

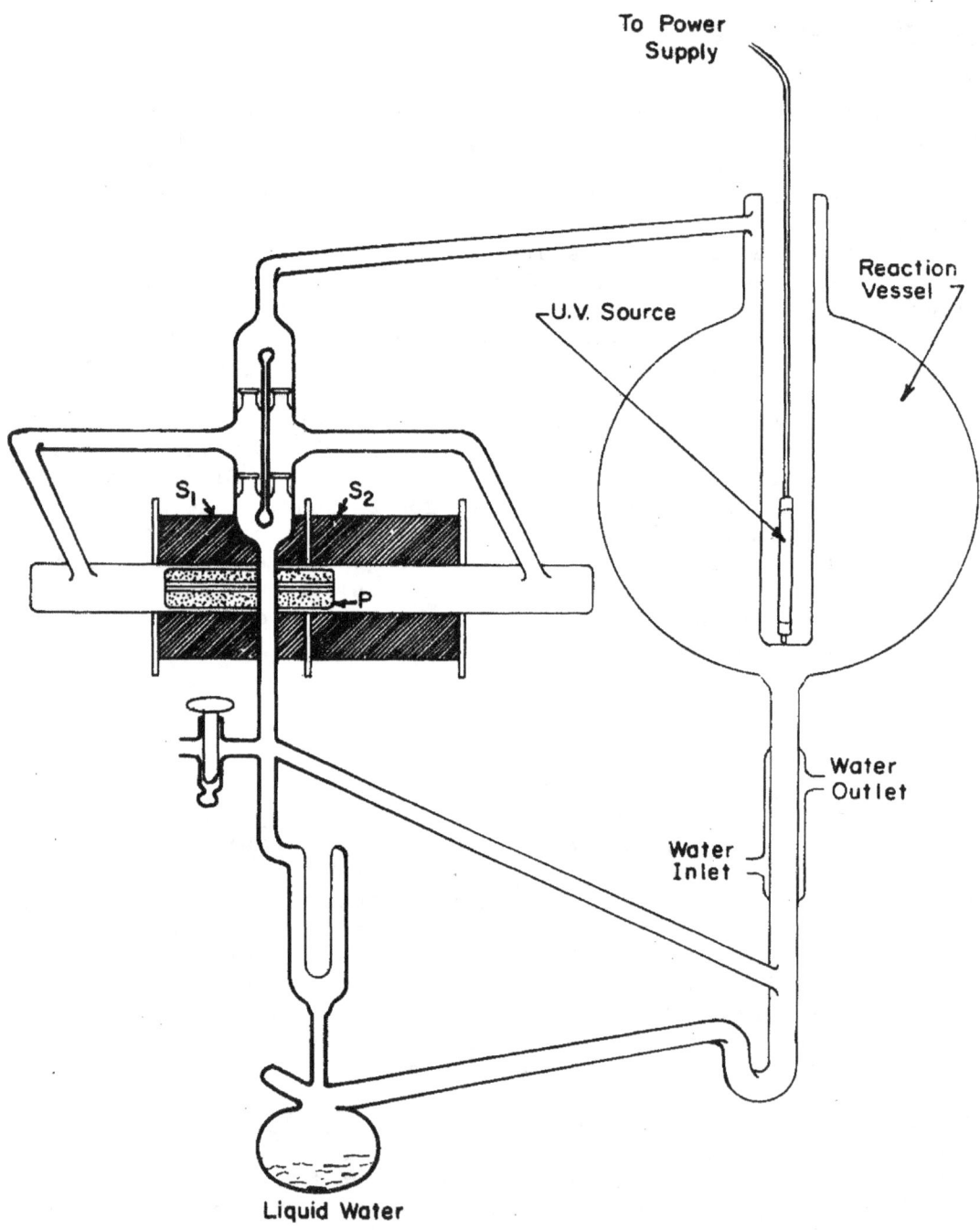

To Power
Supply

Reaction
Vessel

U.V. Source

S_1 S_2

P

Water
Outlet

Water
Inlet

Liquid Water

Figure 2-44.   Schematic illustration of the ultraviolet organic synthetic
experiment at Cornell University. After Khare and Sagan (1973).
Reprinted from Icarus, 20:in press, with permission of
Academic Press, Inc. All rights reserved.

136

Dr. Khare and I have examined the brownish polymer by ultraviolet, visible, and infrared spectroscopy, gas chromatography and mass spectrometry. The preliminary results are that the principal constituent is a long straight-chain hydrocarbon -- an alkane. While polymeric sulfur is made in the aqueous solution it is not a principal constituent of the brown polymer. As far as we can tell, the major chromophores are hydrocarbons. The polymer also contains carbonyl and amino groups, but they do not contribute significantly to the coloration.

Polymeric Transmission Spectra

In Figure 2-45 is the ultraviolet and visible transmission spectrum of the brown polymer. This is a double-beam analysis in which a fragment of the vessel coated with polymer has its transmission spectrum automatically compared with a comparable fragment of a similar reaction vessel without the polymer coating. You can see the very sharp decline in transmission from yellow to ultraviolet, giving this material its reddish coloration. At long visible wavelengths, the transmission is moderately constant. At short optical frequencies the transmission corresponds to something like λ^{-4} and the optical depth to λ^{-2}. There is no reason to think that this is the reddest material that can be produced under these conditions; and indeed we have succeeded in separating via paper chromatography a yellow component of this polymer which runs with the solvent front. This material appears to be substantially redder than the parent brown polymer. I would not at all be surprised if materials with optical depth proportional to λ^{-3} or λ^{-4} are easy to come by in such experiments.

As I've mentioned, the particle size in these experiments is something like 100 μm with a dispersion of a factor of 3 or 4 or 5. The particle sizes are certainly determined in part by wall effects and by the time of irradiation, and I do not know to what extent the 100 μm particle size is characteristic of Titanian conditions. However the results do suggest some caution in concluding that the particles produced in the Titanian clouds must be smaller in diameter than the near infrared wavelengths.

The ultraviolet absorption coefficient is fairly large, about 10^3 cm^{-1} in the near ultraviolet -- which means that a single 100 μm particle has an optical depth in the ultraviolet of about 0.1. It's a pretty sizable absorber. You don't need a lot of this material in order to produce the kinds of optical depths that are talked about in models of the clouds of Titan. Therefore it is entirely possible to imagine a situation in which materials such as this polymer are being made on Titan, fall out in accordance with the Stokes-Cunningham equation, and are replaced by further production -- so that the steady-state abundance is adequate for all optical properties of the clouds.

In Figure 2-46 is an infrared transmission spectrum of two polymeric components -- I is the full brown polymer, and II the yellow chromatographic fraction which I referred to above. There are a number of points of interest: one is the enormous absorption feature at 3 μm which is due to the C-H stretch of the hydrocarbons which are the primary constituent of the polymer.

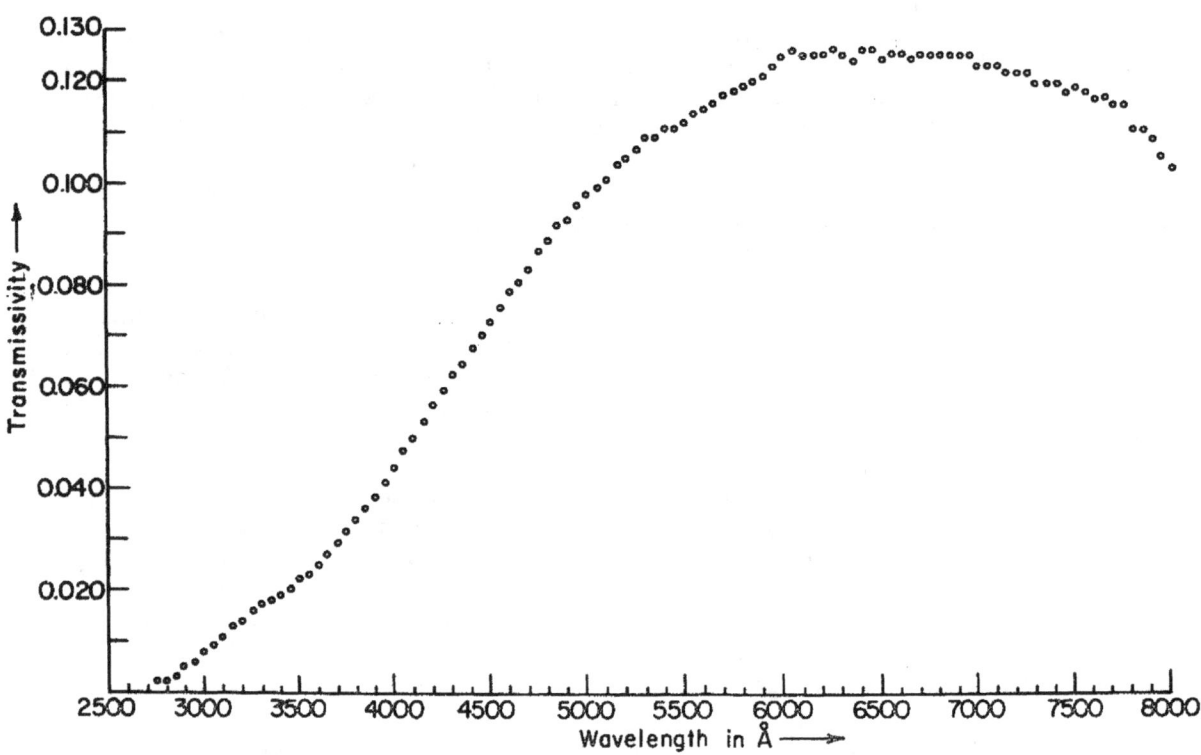

Figure 2-45. Visible and near-UV transmission spectrum of the brown ultraviolet polymer. The spectrum was obtained on a double-beam
spectrometer against a glass sample undercoated by polymer.
After Khare and Sagan (1973). Reprinted from Icarus, 20:in
press, with permission of Academic Press, Inc. All rights
reserved.

138

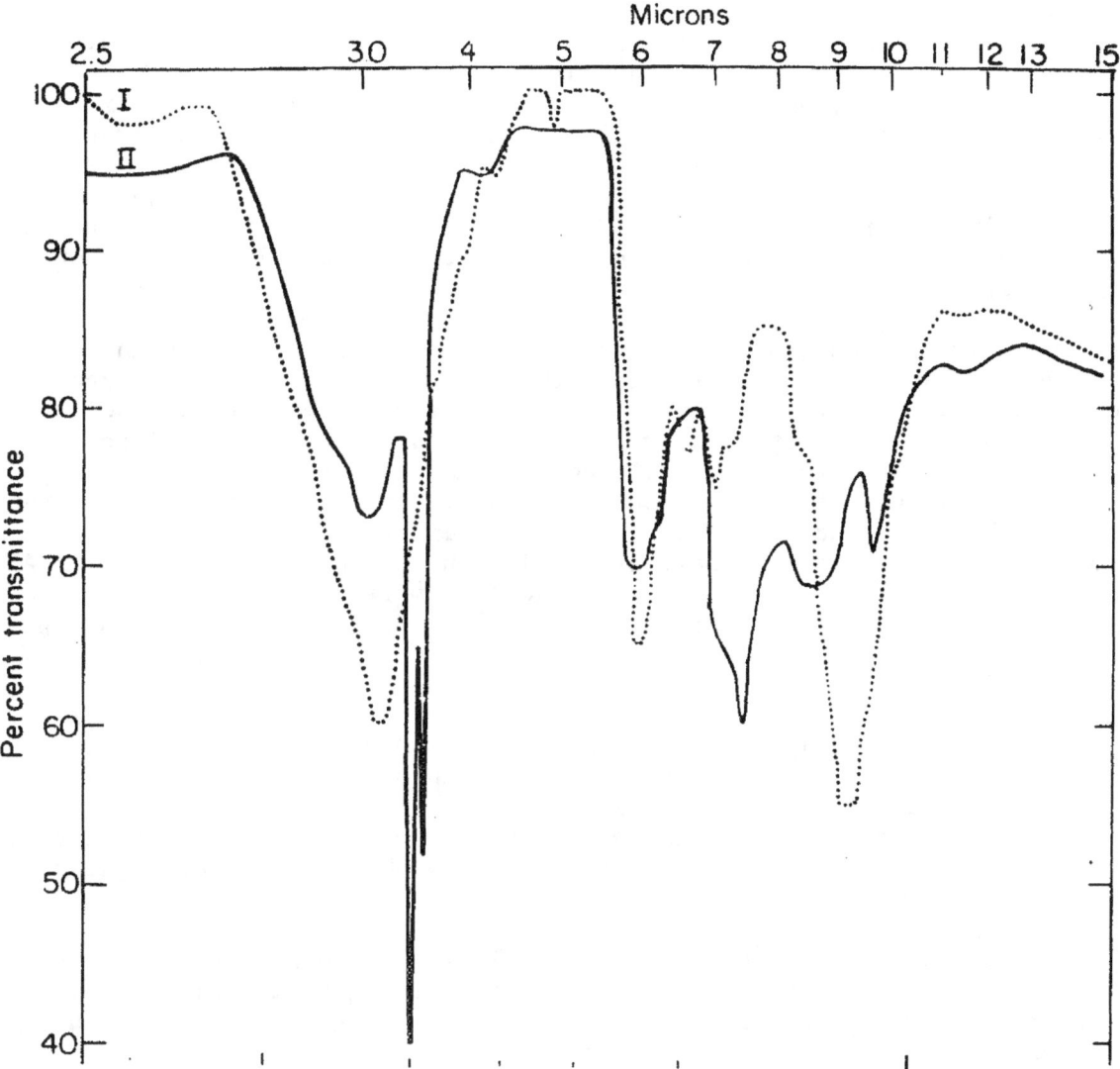

Figure 2-46. Infrared transmission spectrum of the brown polymer and the
yellow-chromatographic fraction of this polymer. After Khare
and Sagan (1973). Reprinted from Icarus, 20:in press, with
permission of Academic Press, Inc. All rights reserved.

139

Next there is substantial and complex absorption in the 7-11 micron region. Therefore it is conceivable that some of the observed absorption features in the 8-13 micron region are due, not to atmospheric emission, but to aerosol absorption. It is also reasonable that the clouds have substantial opacity at 10 μm and that therefore we are not seeing to the surface at such wavelengths.

Another point of interest has to do with the 5-micron window. In material of this sort, the characteristic absorber near 5 μm is the nitrile group, CN. The fact that we are reasonably transparant at these wavelengths means that we are not producing a great many polynitriles. In work by Cyril Ponnamperuma in which his starting materials were ammonia and HCN, he did produce a large quantity of polynitriles and his infrared transmission spectra did show a great deal of absorption at 5 μm. This is of course a reflection of his starting conditions: nitriles in, nitriles out. The results suggest that the examination of the infrared spectrum of such clouds might give some information on the starting conditions.

If our infrared spectrum (Figure 2-46) is typical of the Titanian clouds, then we might see substantially deep into the atmosphere at 5 μm, and experiments of this sort recently reported in a preliminary way by Owen are well worth following up. There is of course multiple scattering in the clouds and even a very large single scattering albedo may imply a small penetration of infrared light through the clouds. But this is true at all infrared wavelengths and the fact that the opacity is lowest in the vicinity of 5 μm suggests that this should be a particularly important wavelength for infrared radiometry. At infrared wavelengths longer than 15 μm I have the impression that there are not a large number of interesting absorption features, but we are still studying this region.

I have talked about very simple products, with as few as three to six atoms, and very complex polymeric material. What about material of intermediate complexity? We do not yet have a good handle on this. We go from simple to complex molecules so fast in such experiments that we are left breathless. Just to give an idea: suppose we take a simple mass spectrum of the polymeric material. We find that every mass-to-charge number from 1 to 1000 is occupied. The complexity of the material is striking.

I have not mentioned much about the chemistry of the aqueous phase in these experiments because there is still a substantial question of whether ammonia hydroxide oceans can exist on Titan. But the organic chemistry in such a liquid medium is of considerable interest.

Conclusions

There is a tradition in astronomy of extreme caution on the question of biology on this or that planet. I believe that Titan looks biologically very promising. There might be as serious an error in being prematurely over-cautious as in being prematurely over-enthusiastic. I do not think we would be doing any disservice if we said that the planning of exobiological experiments on Titan makes some sense. However, even in the absence of any biology on Titan, the presence of interesting organic chemistry -- possibly related to the origin of life on Earth -- looks so promising that I think substantial support from the biological community could be forthcoming in planning, for example, gas chromatograph/mass spectrometer entry probes into the Titanian atmosphere. Apart from the Viking GC/MS which is intended as a post-landing experiment, the gas chromatograph being planned for Venus Pioneer entry probes might be an appropriate precursor instrument.

While the relative hydrogen abundance elsewhere is greater than on Titan, it also seems possible that the Titanian organic chemistry is characteristic of that of the outer solar system in general. Since Titan is the easiest body with an atmosphere to enter in the outer solar system, the exploration of Titan may be the primary stage in the study of the organic chemistry of the entire outer solar system.

Acknowledgement

The recent experimental work reported here has been done jointly with Dr. B. N. Khare, and will be published shortly (Khare and Sagan Icarus, 20). This paper was supported by NASA/JPL Grant NGR 33-010-101.

Pollack: In discussing the laboratory experiments, did you imply that the aldehydes were produced in passing through liquid water?

Sagan: The initial experiments were all in the gaseous phase. However, if a somewhat artificial experiment was performed whereby the products from photolysis or pyrolysis of methane, ammonia and hydrogen were formed in the absence of water vapor, but were then carried into liquid water, the results would be similar. The experiment shown in Figure 2-44 has water in both the vapor and liquid phases.

Danielson: Am I right in saying that you use H_2S as an experimental convenience or are you simulating a Titanian atmosphere?

Sagan: Although there has been some discussion of the possibility of H_2S in the atmospheres of the Jovian planets, and indeed the cosmic abundance of sulfur is high, we have used H_2S as a convenient laboratory source for hot hydrogen atoms. I should stress that none of the color effects which we see in the brown coating are due to polymeric sulfur.

Blamont: I would be happy to provide you with a Lyman-α source which would fit inside your experiment.

Sagan: Thank you; that would be excellent. To substitute Lyman-α excitation for H_2S would be very useful.

Chapter 3

ADDITIONAL CONTRIBUTIONS

3.1   RADIO OBSERVATIONS OF TITAN

       F. H. Briggs

       Saturn and Titan have been observed at three frequencies (1420, 2695, and
8085 MHz) with the NRAO interferometer. As yet, Titan has not been detected.
The observing technique used to separate Titan's signal from that of Saturn
requires long baselines. At the highest frequency, where the signal is strongest and detection most likely, the interferometer "resolution" is so great that
Titan's position must be known to better than a half second of arc throughout
the long integrations needed to reach a low noise level. I have not been able
to obtain ephemerides of this accuracy nor to find an opinion on the accuracy
of positions calculated from the L, M, Θ, and γ tabulated in the American
Ephemeris and Nautical Almanac. At 1420 MHz, where positional accuracy is
not a problem, the signal is much weaker and an upper limit of 1500°K can be
placed on the brightness temperature of Titan.

Postscript, December 3, 1973: A clear positive detection has now been obtained
at 8085 MHz with the NRAO interferometer. Assuming Titan's radius is 2500 km,
the radio brightness temperature (i.e. for unit emissivity) is 115 ± 35°K.
A complete description will be submitted for publication elsewhere.

3.2 TEMPERATURE OF THE THERMOSPHERE

D. F. Strobel

The globally averaged vertical temperature contrast (exospheric temperature minus mesopause temperature) in planetary thermospheres depends on heating by absorption of solar EUV energy, energy loss through infrared radiation by polyatomic molecules and energy transfer by thermal conduction between the regions of energy deposition and loss. On the basis of such a description, Strobel and Smith (1973) estimated a vertical temperature contrast for the thermosphere of Titan $\sim 90°K$ for CH_4/H_2 mixing ratios $\sim 10^{-3}$. Interpretation of current observational data suggests that the CH_4/H_2 mixing ratio $\gtrsim 1$. As a consequence the separation distance between energy deposition and loss is substantially less and we would expect vertical temperature contrasts <10 degrees. Complications do arise however for the thermosphere of Titan. Light constituents such as H_2 can readily escape from its atmosphere and the flow velocities can be large (Hunten 1973). It is quite possible that the light constituents could undergo significant adiabatic cooling and hence have temperatures significantly less than the background atmosphere. In addition, a large relative velocity difference between the light constituent and the background atmosphere can result in considerable energy transfer between the gases and frictional heating of the gases. It is highly probable that H_2 and CH_4 will not be in thermal equilibrium in the thermosphere for large H_2 escape rates.

3.3 8-13 MICRON OBSERVATIONS OF TITAN*

F. C. Gillett, W. J. Forrest, and K. M. Merrill

Abstract

Narrow-band ($\Delta\lambda/\lambda \simeq 0.015$) observations of Titan at selected wavelengths in the 8-13 micron range show evidence for a strong temperature inversion and the existence of at least one more spectroscopically active component in the atmosphere in addition to H_2 and CH_4.

Introduction

Titan has recently been the object of numerous investigations, both observational and theoretical. This surge in interest has been primarily due to the detection of H_2 (Trafton 1972) and the observations of high brightness temperatures in the 8-20 micron range, the latter of which suggest that the atmosphere of Titan may be massive enough to produce a substantial greenhouse effect (Morrison, Cruikshank, and Murphy 1972). This paper describes observations of Titan in the range of 8-13 μm with a resolution $\Delta\lambda/\lambda \simeq 0.015$ and discusses some of the implications of these observations.

Observations

The observations reported here were obtained using a cooled filter-wheel spectrometer together with the UCSD-University of Minnesota 60-inch (152-cm) telescope on Mount Lemmon. The data acquisition and reduction have been discussed elsewhere (Gillett and Forrest 1973). The results are shown in Figure 3-1, where the observed surface brightness is plotted as a function of wavelength, assuming the radius of Titan to be 2.44×10^3 km (Allen 1963). The surface brightnesses are reduced by about 10 percent using the radius proposed for Titan by Morrison et al. (1973). Vertical bars indicate ±1 standard deviation of the mean of measurements at that wavelength, and the horizontal bars indicate the half-power bandpass of the filter used. Also included in this figure are broad-band measurements taken at various times, and curves of constant brightness temperature.

If the actual spectrum of Titan can be approximated by smooth curves joining the spectrometer data points, the 8.4-micron and 12.5-micron broad-band observations appear to be in good agreement with the spectrometer results; however, the 11-micron broad-band measurements seem to be high by a factor of about 1.5. This may indicate the existence of further structure in the spectrum between 10 and 12 μm.

---

* Reprinted from The Astrophys. J., 184:L93-L95, with permission of the University of Chicago Press. © 1973. The American Astronomical Society.

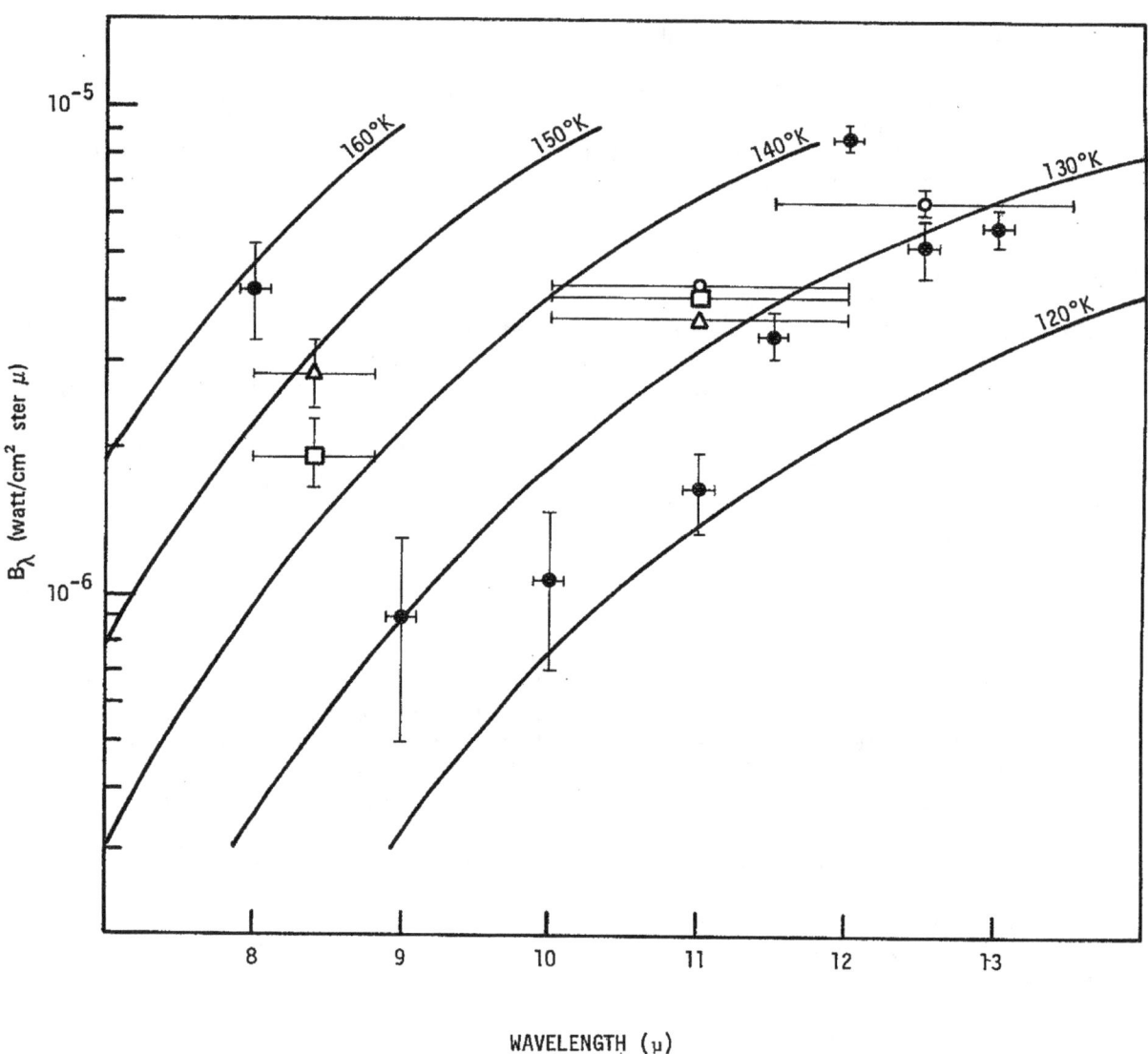

Figure 3-1. Surface brightness of Titan as a function of wavelength. Spectrometer observations: <u>filled circles</u>, 1/1/73 and 2/15/73. Broad-band
observations: <u>open circles</u>, 1/1/73; <u>open triangles</u>, 9/29/72; open
<u>squares</u>, 11/16/71. Horizontal lines indicate bandpass of filter
used.

Discussion

The most significant features of these data are:

(1)   The observations require the presence of at least one spectroscopically active component in addition to the previously detected H_2 and CH_4. If, in the levels where the 9-13 micron radiation originates, the temperature is decreasing with increasing height, then a strong absorber around 10 μm would be required to explain the temperature minimum near this wavelength. Such an absorber could be NH_3 with its ν_2 fundamental near 10.5 μm. The models calculated by Pollack (1973), including the effects of NH_3 absorption, give brightness temperatures in the NH_3 band of about 120°K which is consistent with the observations. On the other hand, if the radiation originates above a temperature inversion (as suggested by Caldwell, Larach, and Danielson 1973), then an additional radiator is required to explain the maximum in brightness temperature around 12 μm (probably C_2H_6). The wavelength coverage and statistical accuracy of the data presented here are not sufficient to rule out either of these possibilities;

(2)   The high brightness temperature within the 7.7-micron CH_4 band shown by the measurement at 8.0 μm definitely indicates the presence of a temperature inversion. A similar elevated brightness temperature within the ν_4 fundamental of CH_4 has been found earlier in the spectrum of Jupiter (Gillett, Low, and Stein 1969), and its association with a temperature inversion was demonstrated by the detection of limb brightening at 7.9 μm (Gillett and Westphal 1973). A surprising aspect of the observations of Titan is the strength of this temperature inversion. The surface brightness of Titan at 8.0 μm is about 6 times that of Jupiter at the same wavelength and about 20 times that of Saturn. The most likely reason for this is that the CH_4/H_2 ratio is much larger for Titan than for Jupiter or Saturn, thus H_2 cooling of the upper atmosphere of Titan is much less effective. Another possibility is that the rate at which energy is absorbed per unit area in the upper atmosphere of Titan may be somewhat higher than for Saturn, depending on the mechanism producing the inversion, but this effect alone cannot account for the difference in the 8.0-micron surface brightness.

For the inversion on Jupiter it was suggested by Gillett et al. (1969) that the energy balance in the inversion layers was determined by absorption of solar radiation in the 3.3-micron band of CH_4 plus possibly some contribution in overtone bands, and radiation through the 7.7-micron CH_4 band and collision-induced transitions in H_2. For Titan it appears that the rate at which energy is being radiated from the upper atmosphere by the 7.7-micron CH_4 band alone, cannot be balanced by absorption of solar radiation via the 3.3-micron and 2.35-micron CH_4 bands. In fact, it is not clear whether the observed inversion could be maintained even if one considered all the overtone bands of CH_4.

A possible alternative source of energy for an inversion that does not suffer from this difficulty is the mechanism proposed by Caldwell et al. (1973), in which solid particles in the upper atmosphere absorb visual and ultraviolet radiation from the Sun and in turn transfer this energy to the gas through collisions. These authors have proposed that the elevated brightness temperatures in the 8-13 micron range reported earlier (Morrison et al. 1972) are due to

147

emission from above a temperature inversion and predicted emission peaks at 7.7 μm (CH$_4$), 12.2 μm (C$_2$H$_6$) and 10.5 μm (C$_2$H$_4$). The data reported here show strong emission peaks in the 7.7-micron CH$_4$ band and at 12 μm. Unfortunately, no measurements were made at 10.5 μm.

Additional insight into the properties of Titan's atmosphere could be obtained from model calculations taking into account the observations reported here. Important additional observational data would include higher spectral resolution observations around 20 μm and better wavelength coverage in the 8-13 micron range.

Acknowledgement

The authors would like to acknowledge discussions with J. Pollack, R. E. Danielson, and D. Morrison. This work was supported under NASA Grant NGL 05-005-003.

3.4 STELLAR OCCULTATIONS

J. Veverka

Introduction

Stellar occultations provide a unique way of obtaining rigorous information on the composition and thermal structure of Titan's upper atmosphere. Although detailed predictions by Taylor (1973) indicate that no suitable occultations by Titan will occur during 1973 and 1974, statistical calculations by O'Leary (1972) predict about 2 "passable" occultations per year, and one "good" occultation every 5 years. A "passable" occultation involves an intensity drop of at least 10% in the U; for a "good" occultation the intensity drop exceeds 50% of the U.

No occultation by Titan has ever been observed photo-electrically. To indicate the kinds of information that potentially can be obtained from occultation light curves, we turn to some observations of the occultation of Beta Scorpii by Jupiter on May 13, 1971.

Atmospheric Structure from an Occultation Light Curve

Simultaneous light curves in three channels (0.353, 0.393, 0.620 μm) of the emersion of Beta Scorpii AB are shown in Figure 3-2 at a time resolution of 0.2 second (Veverka et al. 1973). These light curves can be "inverted" to yield refractivity profiles, and once an atmospheric composition is assumed, temperature and number density profiles such as those shown in Figure 3-3 can be obtained (Wasserman and Veverka 1973b).

An outstanding feature of the light curves shown in Figure 3-2 is the occurrance of numerous light flashes or "spikes", which can be interpreted as small fluctuations in the refractivity profile of the atmosphere. Similar spikes have been observed during an occultation by Neptune (Freeman and Lyngå 1970) suggesting that these fluctuations may be a characteristic of the upper atmospheres of all Jovian planets, and by extension, of Titan's as well. Note that the spikes in Figure 3-2 translate into wiggles in the temperature profile in Figure 3-3.

Atmospheric Composition from Spikes

Whenever high time resolution records of occultation events are obtainable simultaneously in several colors, wavelength dependent time delays in spike arrival times will be observed (Brinkmann 1971; Wasserman and Veverka 1973a). From the differences in spike arrival times at different wavelengths it is possible to determine the relative refractivity of the atmosphere at these wavelengths. Spike arrival delays were observed during the emersion of Beta Scorpii AB, for example, with observations taken every 0.01 second (Veverka et al. 1973). The time delays are well-defined and increase systematically with wavelength. Assuming that the Jovian atmosphere is well

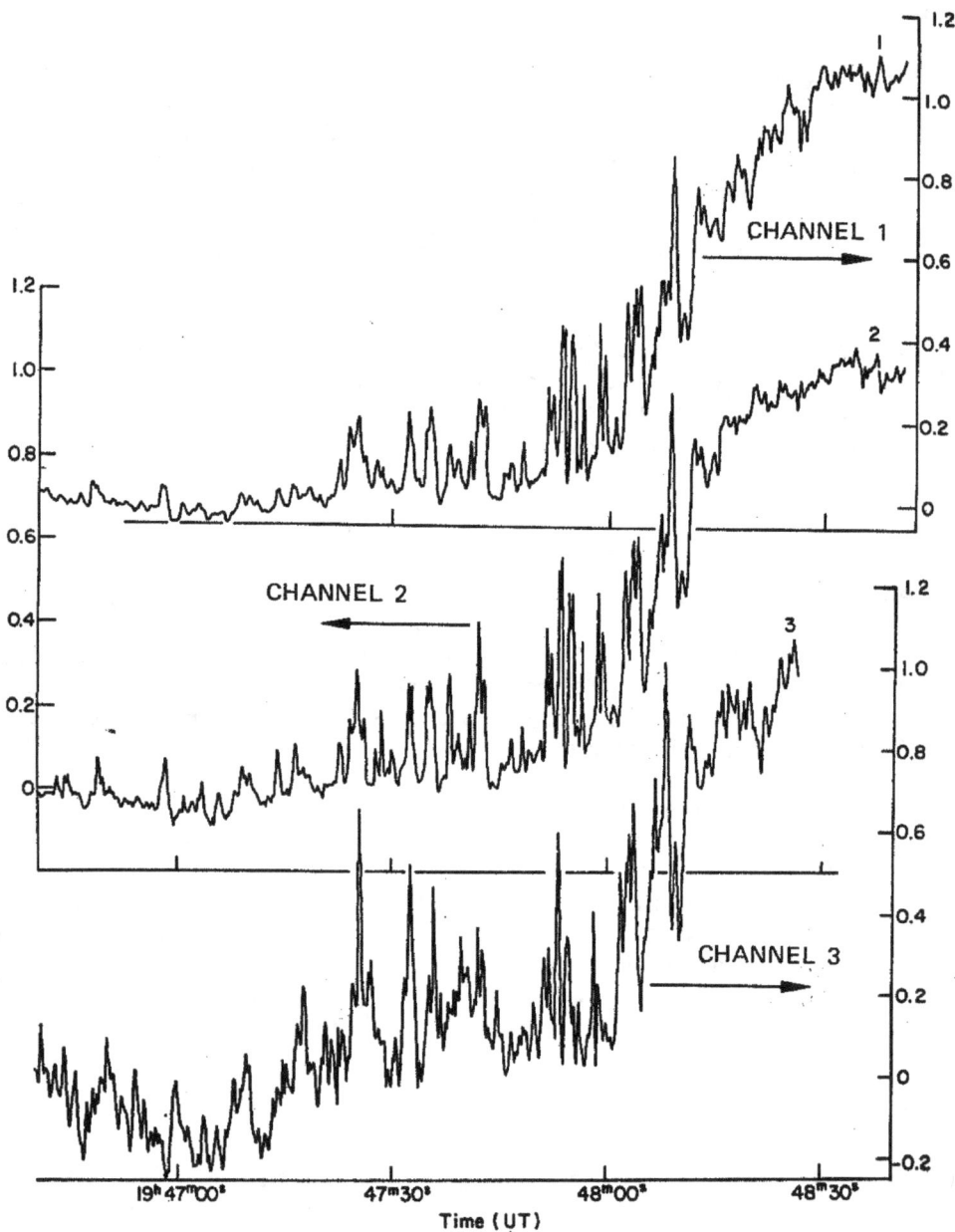

Figure 3-2. Emersion of Beta Scorpii AB in three channels (0.353, 0.393, 0.620 µm) at Δt = 0.2 sec. (Veverka et al. 1973). The zero level base-line was determined by averaging the output over several minutes at a point far removed from the time of emersion. The full scale level was determined by setting the star's full intensity equal to unity. After Veverka, et al. (1974). Reprinted from The Astronomical Journal, in press, with permission of The Astronomical Journal, Dr. L. Woltjer, Ed. All rights reserved.

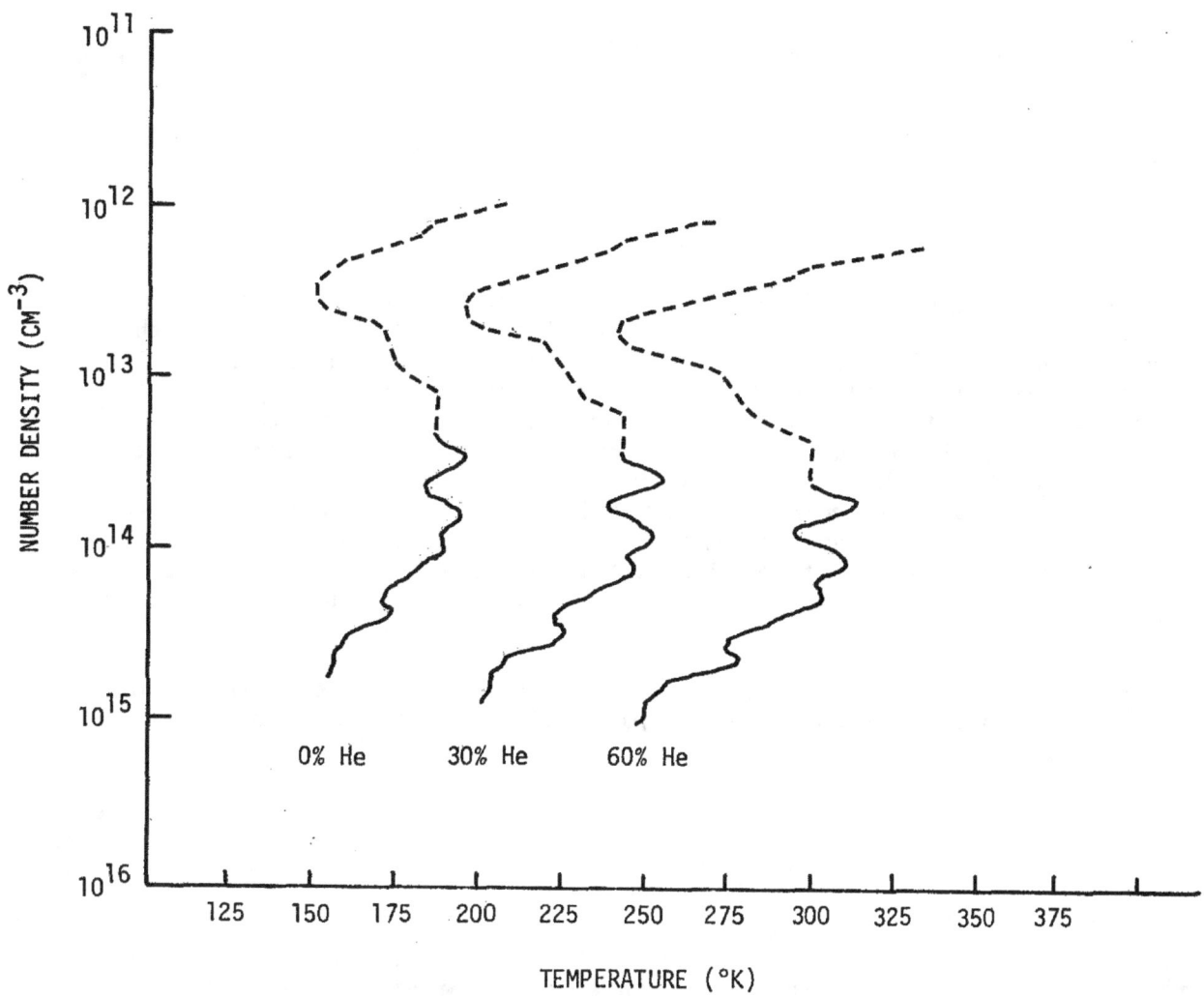

Figure 3-3.   Temperature profiles derived from Channel 2 data for three assumed
              compositions of the Jovian atmosphere. Left to right: 0% He,
              100% H_2; 30% He, 70% H_2; 60% He, 40% H_2 (by number). The upper
              portions of the profiles are uncertain and are therefore shown
              dashed (cf. Wasserman and Veverka 1973b). The bumps correspond
              to the spikes in the light curve. (Although not shown, the pro-
              files derived from the other two channels are similar.) After
              Veverka et al. (1974). Reprinted from The Astronomical Journal,
              in press, with permission of The Astronomical Journal, Dr.
              L. Woltjer, Ed. All rights reserved.

mixed at this level and consists mostly of hydrogen and helium, Elliot et al. (1973) derive the relative abundance of these gases to be:

$$He/H_2 = 0.19^{+.35}_{-.19}$$

by number, which corresponds to a mean molecular weight of $2.32^{+.38}_{-.32}$.

Outlook for Titan

Even if spikes cannot be detected in light curves of occultations by Titan, reliable temperature/number density profiles will be determined if the atmospheric composition is assumed. A strong discrimination among composition models will be possible if an independent estimate of reasonable upper atmosphere temperatures exists. For simplicity, assume that an isothermal structure is found with $T = 100°K$ for pure H_2. Since in this case $T \sim \mu$ (where μ is the mean molecular weight) we have the situation summarized in Table 3-1. It is easy to discriminate against compositions which give high μ's, and hence high temperatures. Thus in this case an atmosphere consisting of 50% of either CH_4 or N_2 could be excluded since it would imply temperatures $>450°K$. However, as Table 3-1 shows, it is difficult to discriminate against high helium fractions in this way.

Fortunately, timing of spike arrival times is most successful in detecting large helium concentrations. For example, it turns out that from the refractivity ratio:

$$\frac{\nu_2}{\nu_3} = \frac{\nu(3934 \text{ Å})}{\nu(6201 \text{ Å})}$$

that large amounts of He are easily detected in the presence of CH_4, N_2 or H_2, but large amounts of N_2 or CH_4 in the presence of H_2 (or vice versa) are not. The situation can be slightly improved by using a larger spread in wavelengths, or more than two wavelengths.

Summary

Stellar occultations can yield reliable temperature/number density profiles of Titan's upper atmosphere near the 10^{14} cm^{-3} number density level. The temperature profiles can be used to discriminate between atmospheric compositions having high mean molecular weights ($\mu > 5$), and those having low mean molecular weights ($\mu < 5$), as indicated in Table 3-1. Detection of spikes at several wavelengths with high time resolution can set useful limits on the helium content of the atmosphere.

Table 3-1. Temperature Dependence on Composition of Hypothetical Upper Atmospheres (See text for details).

CONSTITUENTS	COMPOSITION (BY NUMBER)	MEAN MOLECULAR WEIGHT	T (°K)
H_2 and He	100% H_2	2.0	100
	90% H_2 and 10% He	2.2	110
	50% H_2 and 50% He	3.0	150
H_2 and CH_4	90% H_2 and 10% CH_4	3.4	170
	50% H_2 and 50% CH_4	9.0	450
H_2 and N_2	90% H_2 and 10% N_2	4.6	230
	50% H_2 and 50% N_2	15.0	750
H_2, CH_4, and N_2	80% H_2, 10% CH_4, and 10% N_2	5.0	250
	50% H_2, 25% CH_4, and 25% N_2	12.0	600

It is therefore important that future occultations by Titan be predicted well in advance, and imperative that they be observed adequately. Such observations will also yield an accurate value of Titan's diameter. Accurate diameter determinations can also be made by observing occultations of Titan by the Moon; however, lunar occultations cannot provide any useful information about Titan's atmosphere.

Acknowledgement

I wish to thank C. Sagan, L. Wasserman, J. Elliot and K. Rages for helpful discussions. This work was supported in part by NGR-33-010-082.

3.5   TITAN PHOTOMETRIC PARAMETERS*

R. L. Younkin

Measurements of the irradiance of Titan from 0.50 to 1.08 μm were carried out January 2 and 3, 1972, using the Mount Wilson 60-inch reflector, Fastie-Ebert spectrometer, and pulse counting electronics. An S-20 photomultiplier was used to obtain measurements from 0.50 to 0.795 μm, and an S-1 photomultiplier for measurements from 0.807 to 1.08 μm. A 30 Å exit bandpass was used both nights.

Selection of particular wavelengths of measurement was based upon previous continuous wavelengths energy scans of Jupiter made by the author. With the assumptions that CH_4 is the principal source of absorption in the spectrum of Titan and that the strengths of the absorptions are similar for Titan and Jupiter the wavelengths were chosen first to show the apparent maxima and minima (absorption band centers), and second to obtain values every 100-200 Å.

The comparison standard stars were ξ^2 Ceti, α Leonis, ε Orionis, and γ Geminorum. The values used for the irradiances of these stars as well as the solar irradiances have been given previously by Younkin (1970). The former were based upon relative measurements of Hayes (1967) converted to an absolute scale by the measurements of Willstrop (1960). The latter were based upon measurements of the solar intensity by Labs and Neckel (1967).

The results of reduction of the measurements are given in Table 3-2. A correction to zero solar phase angle of Titan has been made according to the phase law of Blanco and Catalano (1971). The correction increased the brightness of Titan measured here by 0.03 mag.

The irradiances of Titan at the position of the Earth, H, are given in ergs/cm^2/sec/Å and are reduced to mean opposition distances of the Earth and Saturn. The mean surface radiances over the disk of Titan, N, are in ergs/cm^2/steradian/A, corrected to mean opposition distance. For the computation of N and the geometric albedo p_λ (0), Dollfus's (1970) value was used for the angular subtense of Titan at 9.539 AU, 0.700 arc seconds.

The 30 Å bandpass values of the geometric albedo at the measured wavelengths, given in Table 3-2, are plotted in Figure 3-4. An approximate contour has been drawn between the points, with some weight given to the shape of the methane bands of Jupiter, based upon unpublished continuous scans of the center of the disk by the author. This contour thus assumes methane is the principal absorber in this spectral range.

It is seen from Figure 3-4 that the geometric albedo of Titan reaches a maximum value of 0.37 at 0.68, 0.753, and 0.83 μm, while in the center of the strongest absorption features, at 0.89 and 1.01 μm, it falls to 0.10. Comparison with the methane bands of Jupiter and Saturn given below indicates the

---

* Reprinted in part from Icarus, 21:in press, with permission of Academic Press, Inc. All rights reserved.

Table 3-2. Photometric Parameters for Titan

λ	$H \times 10^{12}$	N	$p_\lambda (0)$
0.5000	1.54	0.133	0.209
0.5124	1.65	0.143	0.223
0.5264	1.74	0.150	0.242
0.5370	1.90	0.164	0.260
0.5490	1.97	0.170	0.270
0.5556	2.00	0.172	0.278
0.5652	2.07	0.179	0.290
0.5840	2.17	0.188	0.302
0.5914	2.15	0.185	0.311
0.6024	2.15	0.186	0.317
0.6100	2.15	0.186	0.318
0.6140	1.97	0.170	0.296
0.6190	1.85	0.160	0.279
0.6240	1.98	0.171	0.299
0.6280	2.20	0.190	0.342
0.6370	2.22	0.191	0.343
0.6424	2.19	0.189	0.347
0.6500	2.17	0.187	0.351
0.6630	2.03	0.175	0.331
0.6730	2.07	0.179	0.352
0.6800	2.11	0.182	0.368
0.7020	1.66	0.143	0.305
0.7124	1.88	0.162	0.351
0.7274	1.07	0.093	0.206
0.7364	1.44	0.124	0.280
0.7450	1.81	0.156	0.360
0.7500	1.81	0.156	0.364
0.7530	1.81	0.156	0.368
0.7724	1.53	0.132	0.324
0.7834	1.22	0.104	0.263
0.7890	1.23	0.108	0.263
0.7950	1.12	0.096	0.248
0.8070	1.20	0.104	0.274

Table 3-2. Photometric Parameters for Titan (Contd)

λ	$H \times 10^{12}$	N	$p_\lambda (0)$
0.8172	1.51	0.130	0.353
0.8300	1.55	0.133	0.372
0.8400	0.95	0.082	0.234
0.8470	1.10	0.095	0.276
0.8600	0.69	0.060	0.176
0.8640	0.67	0.058	0.170
0.8670	0.59	0.050	0.150
0.8700	0.69	0.060	0.178
0.8804	0.56	0.048	0.147
0.8872	0.38	0.032	0.101
0.8926	0.38	0.033	0.104
0.9100	0.76	0.066	0.220
0.9358	1.07	0.092	0.317
0.9700	0.43	0.037	0.139
0.9880	0.35	0.030	0.116
1.0120	0.30	0.026	0.105
1.0200	0.36	0.031	0.128
1.0400	0.46	0.040	0.169
1.0600	0.62	0.054	0.246
1.0800	0.78	0.068	0.320

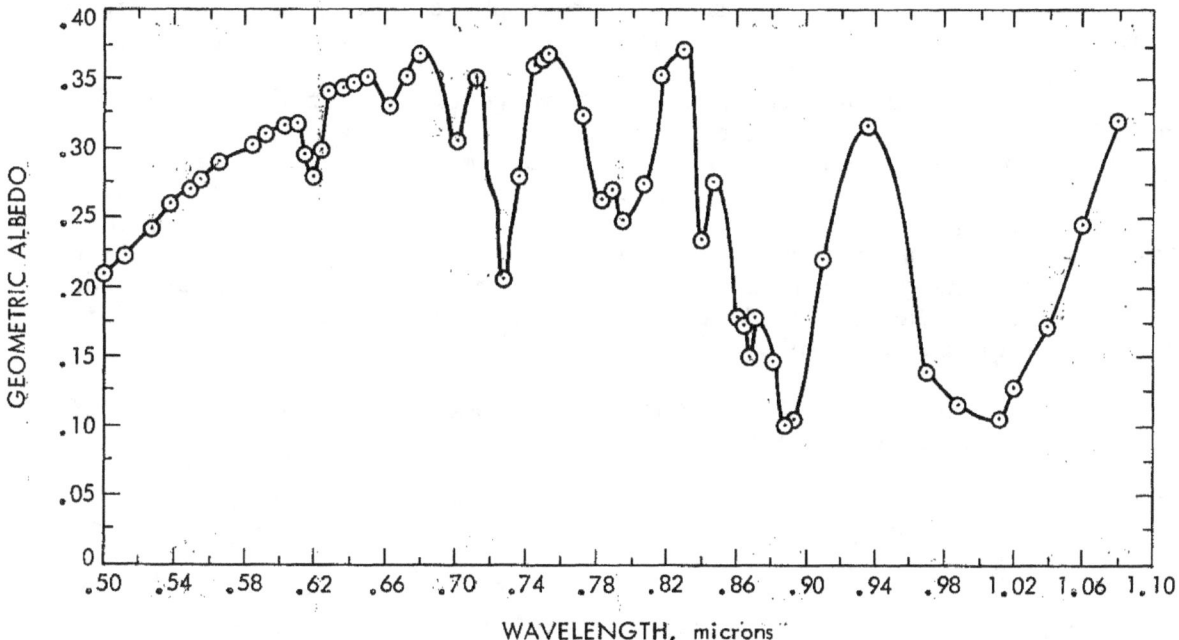

Figure 3-4.   Narrow band geometric albedo of Titan, adjusted to zero planetary
phase angle.   The circles represent measured points, the curve,
the estimated albedo between the points.

maxima at 0.936 and 1.08 μm clearly do not represent the continuum. The points at 0.753 and 0.83 μm may well also be depressed from continuum values by the wings of the methane bands adjacent to them. It is not clear from these measurements whether the "red" continuum short of 0.68 μm flattens out at longer wavelengths.

The bolometric albedo, A*, of a planet or satellite is defined as the ratio of the total reflected flux at all wavelengths to the total solar flux intercepted by the surface. This may be reduced to the well-known expression,

$$A^* = \int p_\lambda \, q_\lambda \, H_\lambda \, d\lambda \Big/ \int H_\lambda \, d\lambda \qquad (1)$$

where p_λ is the geometric albedo, q_λ the planetary phase function, and H_λ the solar irradiance at one AU.

To compute a value for A*, the wavelength range of p_λ must be extended. For 0.3 to 0.5 μm, the medium band measurements of McCord et al. (1971) have been fitted to the present results. The long wavelength limit for pλ has been extended to 4.0 μm to include 98% of the incident solar flux. A hypothetical curve from 1.0 to 4.0 μm has been computed, based upon measurements of the energy from Jupiter and Saturn in this region by Moroz (1966), Danielson (1966), Gillett et al. (1969), and Johnson (1970). It assumes that methane is the principal absorber in this region. The curve adopted for the geometric albedo from 0.30 to 4.0 μm is shown in Figure 3-5.

Assuming a constant value of q_λ so it can be removed from the integral of Equation 1, evaluation of the integral gives

$$A^*/\bar{q} = 0.21 \qquad (2)$$

In the region beyond 1.1 μm where the values of p_λ have been estimated, the solar irradiance is falling rapidly from its maximum near 0.50 μm, so the values adopted for p_λ are not critical. It is believed therefore the value of $A^*/\bar{q}$ given above should be correct to ±10%, subject only to possible changes in the radius of Titan.

From measured values of q for the Earth, Venus, and Mars, as well as calculated values for various particle scattering functions and single scattering albedos, it is suggested the value of $\bar{q}$ for Titan is in the range 1.1 to 1.5, with a most plausible value of 1.3. Admittedly this represents merely an educated guess, but it is believed to be the best that can be made at this time. These values give 0.23, 0.27, and 0.31 for A*. Then the effective radiative temperature of Titan may be computed by the equation

$$T_e^4 = S_o \, (1 - A^*)/D^2\sigma \qquad (3)$$

where S_o is the solar constant, D, the Sun-Saturn distance, and σ, the Stefan-Boltzmann constant. With the values of A* above, $T_e = 84 \pm 2°K$.

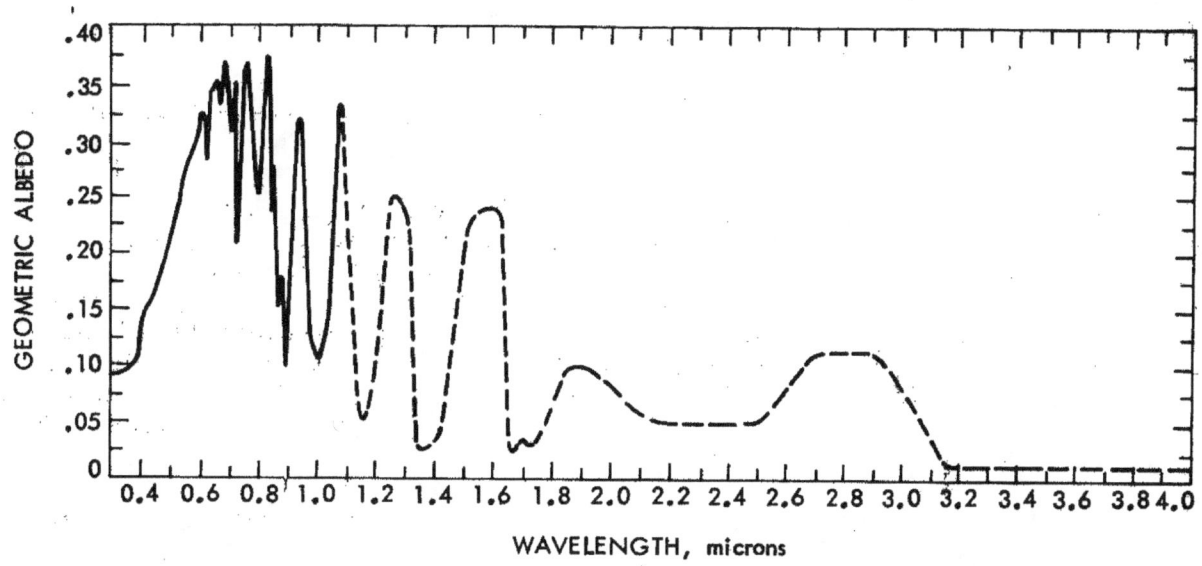

Figure 3-5. Geometric albedo of Titan from 0.3 to 4.0 μm. The solid line represents measured values, the dashed line estimated values.

Acknowledgements

This paper presents the results of one phase of research carried out at the Jet Propulsion Laboratory, California Institute of Technology, under Contract Number NAS7-100, sponsored by the National Aeronautics and Space Administration.

Chapter 4

RECOMMENDATIONS

The recommendations given here were discussed at the Workshop and represent the collective opinion of its members. A surprisingly large number of
actions were identified that can contribute, by existing Earth-based techniques,
to a better definition of the Titanian environment. The corresponding recommendations are prefaced by an asterisk (*). In a time as short as another year,
we can expect much better assurance in our models of Titan's atmosphere. The
remaining suggestions are also scientifically important, but do not meet the
narrow qualification for an asterisk: will it help the engineers and others
who must plan the in-situ exploration of Titan? In most cases, the justification of a statement is not given in this chapter, but can be found in Chapters
2 and 3.

*Infrared Spectrophotometry

Spectrophotometric observations of Titan in the thermal infrared offer
immediate propects of distinguishing among physical models for the atmosphere.
Two spectral regions are of particular interest. Between 15 and 100 μm, most
of the solar radiation absorbed by Titan is re-emitted, so that observation
in this part of the spectrum will be of fundamental importance for an understanding of the most important opacity sources in the atmosphere. Since the
opacity should vary only slowly with wavelength, moderate spectral resolution
$\lambda/\Delta\lambda \simeq 20$ should be sufficient. In the region 17 to 28 μm, these observations
can be made using existing equipment at high-altitudes sites where the low
humidity results in high atmospheric transmission. At such sites, it may also
be possible to detect Titan in a broad band centered at 35 μm. However, spectrophotometry over the entire 28-40 micron band requires that observations be
made from high-flying aircraft. The NASA C-141 will be ideally suited for
these observations.

At wavelengths from 5 to 15 μm, Titan emits radiation at a level of intensity several orders of magnitude greater than would be expected for an object
of its effective temperature. The radiation observed in the 8-14 micron region
has been interpreted as resulting from either a greenhouse effect, and resulting
high surface temperature, or from emission from dust and direct line radiation
from a hot region in the high atmosphere. In either case, there is expected to
be substantial spectral structure in this wavelength band, with the detailed
shape of the spectra diagnostic of the temperature structure of the atmosphere.
Present facilities are capable of spectral resolution $\lambda/\Delta\lambda = 100$, and spectrophotometry with this resolution throughout the region 8-14 μm would be extremely
useful.

We recommend that support be made immediately available to observers who have the capability of obtaining thermal emission spectra with resolution of 100 between 8 and 14 μm and with resolution 10 to 20 between 16 and 28 μm. Such experiments can provide diagnostic data within the next 12 months. We further recommend that support be given to plans to use the NASA C-141 airborne telescope facility at the earliest opportunity to obtain spectra of Titan out to 100 μm.

*Stellar and Lunar Occultations

A strong effort should be made to observe all occultations of stars brighter than magnitude +10, by Titan during the next five years. On statistical grounds about 10 useful occultations during this period are expected.

These observations certainly will yield accurate information about the density and temperature structure of Titan's upper atmosphere near the 10^{14} cm^{-3} level, and probably, reliable information about the atmospheric composition. In addition, such observations will provide accurate values of Titan's diameter.

To implement this program, reliable predictions of stellar occultations by Titan are needed. Work should be carried out to improve: (a) the ephemeris of Titan; (b) the knowledge of star positions (down to mag +11) in Titan's path. Existing prediction procedures must be improved to make sure that predictions of events are accurate, and are publicized well in advance. An improved ephemeris of Titan should also make it possible to detect Titan at radio frequencies.

Lunar occultations of Titan should be observed. There are about half a dozen such opportunities during 1973-74. Such observations will give an accurate diameter for Titan, but cannot provide useful information about Titan's atmosphere.

*Radio Interferometry

Perhaps the most critical undetermined parameter about Titan is the surface temperature. As for Venus, another cloud-covered solar-system object, microwave observations may be able to determine the surface temperature. For all plausible compositions -- except very high temperatures and very high NH_3 abundances -- the atmosphere and clouds should be transparent at microwave frequencies.

Titan is a difficult radio source -- of small angular size and possibly of low brightness temperature. Its detection is confusion-limited: distant radio sources can be confused with Titan in the large angular-size beam of a single radio telescope. However, precision angular positions can be determined with a radio interferometer. Several adequate interferometers now exist, and an attempt to measure the brightness temperature of Titan at 21 cm with the NRAO interferometer has been made by Briggs and Drake of Cornell. Their work has not been successful primarily because of a lack of accurate knowledge of the position of Titan. Thus, precise ephemerides of Titan may lead to a discrimination among the contending models of the Titanian environment. Detection of brightness temperatures in the 150°-200°K would then be a distinct possibility. We strongly recommend a concerted investigation of the ephemeris of Titan.

*Laboratory Spectroscopy

We recommend that laboratory and theoretical investigations be undertaken to enable the determination of the structure of bands in the 7-14 μm region of molecules which may contribute to the observed Titan emission, e.g., CH_4 (7.7 μm), C_2H_6 (12.2 μm), C_2H_4 (10.5 μm), and C_2H_2 (13.7 μm). Other molecules with permitted bands in this region can probably be ruled out on various chemical grounds. Furthermore, investigations of the broadening of the pressure-induced pure rotational and translational transition of H_2, induced by CH_4 and N_2, especially on the short wavelength wing, would be desirable. The temperature dependence of the structure of these bands should be included in these studies.

The interpretation of a wealth of current spectroscopic data at shorter wavelengths is severely limited by a lack of laboratory measurements of various gases.

In order to identify the gases in Titan's atmosphere that are responsible for absorptions not visible in the spectrum of Saturn, spectra of a wide variety of plausible gases need to be obtained in the laboratory. These should be obtained under conditions which permit the form of these spectra to be derived for the physical conditions existing in Titan's atmosphere. Such comparison may not only permit the identification of an unknown gaseous constituent but also shed some light on its relative abundance and environmental pressure. If such constituents turn out to contain rare isotopes, important isotopic ratios may be obtained which are of value to studies of atmospheric evolution.

Laboratory spectra are required of $^{12}CH_4$, $^{13}CH_4$, CH_3D, higher hydrocarbons and their isotopes, and likely products of the photolysis of CH_4 for the identification of unknown features in Titan's spectrum. These should preferably be obtained at reduced temperature (100°K) to facilitate matching with Titan's spectrum. Resolution elements of 0.1 Å are required for wavelengths between 0.6 and 1 μm and 0.5 Å or better between 1 and 2 μm.

Line-broadening coefficients for the $3\nu_3$ CH_4 band are required at low temperatures for self-broadening and broadening by H_2 and N_2.

Low-resolution spectra (30 Å) are required as a function of pressure and path length for the CH_4 bands between 6000 Å and 2 μm at Titanian temperatures to establish curves of growth of these bands and of typical lines in parts of these bands. For these studies, low resolution is sufficient.

Corresponding curves of growth should be obtained for CH_4 diluted in N_2 and H_2 as well. The data should clearly show the transition from the linear to the square-root regime. For CH_4, path lengths of CH_4 as long as 2 km-A are required for Titan.

*Near-Infrared Spectroscopy

Practically all our present knowledge of Titan's atmospheric composition comes from spectroscopy in the 7000-11000 Å region. Many additional observed absorptions are not understood. Important information can be expected from work at higher resolution and extensions to longer wavelengths. Such work should be pursued to the limit of available telescope time.

Potential Clouds

We recommend that laboratory optical studies of potential cloud-forming materials be conducted. Among the most important materials are liquid and solid methane, solid ethane, ethylene, and propane, and photochemically produced polymers from solar UV irradiation of methane. The most urgent properties to be measured are the UV, visible, and IR reflection spectra and the complex refractive index of these particles.

We also recommend: (a) that laboratory experiments be expanded to produce as realistic as possible samples of the dust which is postulated to occur in the atmosphere of Titan; and (b) measurements be made of the complex index of refraction, composition, and other physical properties of the samples produced.

The Hydrogen Toroid

Hydrogen escaping from Titan accumulates in the region of Titan's orbit and should form a toroidal cloud around Saturn whose angular dimension may be 10 arc min. A small part of this cloud is atomic hydrogen, which is easily excited by solar light to the first excited states. Preliminary computations indicate a flux of the order of magnitude of 500 Rayleighs in Lyman-α light in the region of maximum emissions, to be compared with the flux due to interplanetary hydrogen (200 to 500 Rayleighs) and to geocoronal hydrogen (2000 Rayleighs during the night at 150 km altitude); the detection of the cloud appears therefore possible.

Measurement of the Lyman-α flux of such a possible cloud would definitely demonstrate the presence of H_2 in the atmosphere of Titan and provide a quasi-direct way of determining its density distribution.

Physics of the Interior

We recommend that high-pressure equation-of-state and phase-stability studies of ices in the H_2O-NH_3-CH_4 system be conducted. It is most important to span the temperature range from 60 to 400°K and pressures up to 10 kb. Realistic internal static and thermal structure models require the availability of such data, and inference of the internal composition of ice-rock satellites is impossible without it. It is our expectation that existing laboratory equipment would be suitable to this purpose.

Organic Synthesis in Simulated Atmospheres

Experiments should be performed in which the environment, atmosphere and clouds of Titan are simulated, appropriate energy sources employed, and the resulting organic and other compounds analyzed. Useful work can be done at room temperature and 1 bar pressure, but experiments at temperatures down to 77°K would be useful. Laboratory energy sources of interest include hydrogen Lyman-α, long wavelength ultraviolet light, electrical discharges and hypervelocity shocks.

Qualitative work, as on the gas chromatograph/mass spectrometry of products, as well as quantitative work, as on the quantum yields of simple organics, would both be useful. A range of progressively more complex precursors, beginning with hydrogen and methane alone, should be employed. The ultraviolet, visible, infrared, and microwave properties of the products should be investigated to compare with observations of Titan. Similar work is presently being funded by NASA in the context of experiments related to the origin of life; but experiments more precisely connected to the actual Titanian conditions should be encouraged.

Sterilization of Titan Entry Probes

Present evidence suggests that Titan's atmosphere may contain large quantities of organic molecules and that its surface may be much warmer than expected from its albedo and heliocentric distance. These are conditions almost certainly consistent with survival, and possibly consistent with growth, of terrestrial anaerobic heterotrophic or photoautotrophic micro-organisms. Since eventually spacecraft investigations of Titan will include exobiological experiments, it is important to prevent contamination of Titan by terrestrial micro-organisms inadvertently carried by unsterilized spacecraft. Accordingly Titan entry probes should be scrupulously sterilized -- at least until our knowledge of the Titanian environment is significantly improved.

The logic is the same as for sterilization of Mars entry probes, which has been agreed to by both the United States and the Soviet Union and carried out on the Mars 2 and 3 space vehicles. A sterilization protocol for Titan-bound spacecraft comparable to that for Mars-bound spacecraft has been recommended by the Panel on Planetary Quarantine of COSPAR, the Committee on Space Research of the International Council of Scientific Unions. Instruments for Titan entry probes should be designed and selected in light of the need for sterilization.

Chapter 5

NASA MISSION PLANNING

This chapter discusses the conclusions of the Workshop relative to the Mariner Jupiter/Saturn (MJS) missions and current studies of possible Titan entry probes.

MJS Missions

The MJS mission carries the following experiments that could be relevant to Titan:

- Imaging

- Ultraviolet Spectrometer

- Infrared Spectrometer

- Radio Science

In addition, the magnetometer and particle detectors would be of interest if Titan has a magnetic field. A second ultraviolet photometer (the Blamont experiment) may be carried; it is specialized for work on Lyman-α.

We found no reason to consider the MJS flyby a prerequisite to the planning of a probe mission to Titan; sufficient understanding can be acquired by Earth-based observations over the next year or so to make a satisfactory probe recommendation. On the other hand, as we have described above, Titan holds a unique and important position in our studies of planetary evolution and we recognize that the MJS mission provides an important opportunity to examine Titan and its relation to the Saturnian and Jovian systems. For this reason we recommend that the MJS mission planners place great emphasis on selecting trajectories that encounter Titan in such a manner as to guarantee maximum science return from that object.

Hydrogen abundance and its ratio to H_2, CH_4, He, D, in planetary atmospheres is a primary consideration in theories of the formation of the atmosphere as well as the nature of the planet itself. Many implications are drawn from each of the H Lyman-α signatures: emission line-shape, resonance intensity, spatial intensity distribution. The proposed Blamont experiment is a specialized Lyman-α instrument which would be complementary to the selected MJS payload. The high spatial and spectral resolution of this instrument, 1 second of arc and 0.01 Å respectively, would be particularly productive in the exploration of the atmospheres of Titan and Saturn. It could also be crucial in determining the nature of the toroidal clouds of H_2 and H that may be present around the orbit of Titan and possibly other satellites. We therefore recommend that, if at all possible, the Blamont experiment be included in the MJS payload.

Titan Entry Probes

Titan is clearly a prime candidate for an entry mission: the substantial atmosphere and low entry velocity make the job much easier than for Mars (once the vicinity of Saturn has been reached). There are many important questions that are difficult or impossible to attack by remote sensing, but that yield readily to direct measurement. Biological sterilization will be required, even though Titanian temperatures are probably far too low for the growth of terrestrial organisms.

Current advanced planning for probes in the outer solar system visualizes a 3-mission set launched in the early 1980's, aimed at Saturn and Uranus with one combined backup. In principle, this SU probe mission could be augmented to an SUT. An absolute minimum SUT program could perhaps use a third spacecraft for Titan. It would seem much more reasonable to add more spacecraft when the number of targets is increased from 2 to 3. A total of 5 launches would be appropriate.

A second question is whether a common SUT probe design is desirable or even feasible. There is little doubt that a probe designed to enter Saturn's atmosphere can also enter Titan's in safety. However, we are not yet sure that Titan's atmosphere is deep enough to leave an attractive mission after entry. Moreover, if Titan probes must be sterilized, the same burden would have to be laid on at least some of the probes destined for Saturn and Uranus. Finally, as discussed below, the instruments appropriate for preliminary exploration of Saturn and Uranus may not be adequate for Titan.

The Workshop's preliminary conclusion was that a common SUT probe design is not clearly desirable. This suggestion does not rule out a very high degree of commonality; it merely says that a mission set of 5 identical probes for 3 objects does not look attractive at present. Three probes destined from the beginning for SU, and two for Titan, make more sense. However, definite conclusions are premature at present. It would be appropriate for another group to study the issue in more detail in 12-18 months, when much more information about Titan can be expected. This group would require information about mission opportunities and constraints, as well as about Titan.

Studies of SU probes have so far concentrated on a "minimum" scientific payload, appropriate for diagnosing the atmosphere of a Jovian planet consisting mostly of hydrogen and helium. The instruments are:

- Temperature Gauge

- Pressure Gauge

- Accelerometer

- Mass Spectrometer

- Nephelometer

For Titan, with a large abundance of methane and probable photolysis products, a gas chromatograph should supplement (or perhaps even replace) the mass spectrometer. This instrument is part of the preliminary Pioneer Venus payload. Another Pioneer Venus instrument that should be considered is the cloud-particle size spectrometer, which would replace the nephelometer. All present indication is that aerosols and clouds are an important aspect of Titan's atmosphere, and the crude information from a nephelometer is probably not enough.

BIBLIOGRAPHY AND REFERENCES

Allen, C. W., 1963: Astrophysical Quantities, London, Athlone Press.

Allen, D. A., and Murdock, T. L., 1971: Infrared photometry of Saturn, Titan, and the rings. Icarus, 14, 1-2.

Arking, A., and Potter, J., 1968: The phase curve of Venus and the nature of its clouds. J. Atmos. Sci., 25, 617.

Axel, L., 1972: Inhomogeneous models of the atmosphere of Jupiter. Ap. J., 173, 451-468.

Barker, E. S. and Trafton, L. M., 1973: The reflectivity of Titan from 3000-4350Å. Bull. A.A.S., 5, 305.

Barker, E. S., and Trafton, L. M., 1973: Ultraviolet reflectivity and geometrical albedo of Titan. Submitted to Icarus.

Blanco, C., and Catalano, S., 1971: Photoelectric observations of Saturn satellites Rhea and Titan. Astron. and Astrophys., 14, 43-47.

Brinkmann, R. T., 1971: Occultation by Jupiter. Nature, 230, 515.

Brouwer, D., and Clemence, G. M., 1961: Orbits and Masses of planets and satellites, in Planets and Satellites. G. P. Kuiper and B. M. Middlehurst, Eds., Chicago, University of Chicago Press, 31-94.

Caldwell, J. J., 1973: In preparation.

Caldwell, J. J., Larach, D. R., and Danielson, R. E., 1973: The continuum albedo of Titan. Bull. A.A.S., 5, 305.

Carlson, R. W., Bhattacharyya, J. C., Smith, B. A., Johnson, T. V., Hidayat, B., Smith, S. A., Taylor, G. E., O'Leary, B. T., and Brinkmann, R. T.: 1973, An Atmosphere on Ganymede from Its Occultation of SAO186800 on 7 June 1972, Science 182, 53.

Cess, R., and Owen, T., 1973: Titan: The effect of noble gases on an atmospheric greenhouse. Nature, 244, 272.

Coffeen, D. L., and Hansen, J. E., 1973: Polarization studies of planetary atmospheres, in Planets, Stars, and Nebulae studied with Photopolarimetry. T. Gehrels, Ed., Tucson, University of Arizona Press.

Comas Solá, J., 1908: Observations des satellites principaux de Jupiter et de Titan. Astron. Nachr., 179, no. 4290, 289-290.

Cruikshank, D. P., and Morrison, D., 1972: Titan and its atmosphere. Sky and Telescope, 44, 83-85.

Danielson, R. E., 1966: The infrared spectrum of Jupiter. Ap. J., 143, 949.

Danielson, R. E., Caldwell, J. J., and Larach, D. R., 1973: An inversion in the atmosphere of Titan. Icarus, 20 (4).

Delsemme, A. H., and Miller, D. C., 1970: Physio-chemical phenomena in comets. II. Gas absorption in the snows of the nucleus. Planet. Space Sci., 18, 717.

Dennefeld, 1973: Ph.D thesis, University of Paris.

Divine, N., 1973: Titan Atmospheric Models (1973), to be issued by JPL.

Dollfus, A., 1970: Diamètres des planètes et satellites, in Surfaces and Interiors of Planets and Satellites. A. Dollfus, Ed., London, Academic Press, p. 46.

Eichelberger, W. S., 1911: The mass of Titan. Pub. U.S. Naval Obs., 2nd ser., 6, B5.

Elliot, J., Wasserman, L., Veverka, J., Sagan, C., and Liller, W., 1973: The occultation of Beta Scorpii by Jupiter. II. The hydrogen-helium abundance in the Jovian atmosphere. Submitted to Ap. J..

Evans, D. C., 1965: Ultraviolet reflectivity of Mars. Science, 149, 969.

Franklin, F. A., and Cook, A. F., II, 1969: A search for an atmosphere enveloping Saturn's ring. Icarus, 10, 417.

Freeman, K. C., and Lyngå, G., 1970: Data for Neptune from occultation observations. Ap. J., 160, 767.

Gillett, F. C., and Forrest, W. J., 1973: Spectra of the Becklin-Neugebauer point source and the Kleinmann-Low nebula from 2.8 to 13.5 microns. Ap. J., 179, 483.

Gillett, F. C., Forrest, W. J., and Merrill, K. M., 1973: 8-13 μm observations of Titan. Ap. J. Letters, 184, 93-95.

Gillett, F. C., Low, F. J., and Stein, W. A., 1969: The 2.8-14 micron spectrum of Jupiter. Ap. J., 157, 925-934.

Gillett, F. C., and Westphal, J. A., 1973: Observations of 7.9-micron limb brightening of Jupiter. Ap. J. Letters, 179, 153-154.

Goody, R. M., 1964: Atmospheric Radiation. I. Theoretical Basis. Oxford, Clarendon Press, Chap. 4.

Gross, S. H., 1973: The atmosphere of Titan and the Galilean satellites. Submitted to J. Atmos. Sci..

Gross, S. H., and Mumma, M. J., 1973: Private communications.

Harris, D. L., 1961: Photometry and colorimetry of planets and satellites, in Planets and Satellites. G. P. Kuiper, and B. M. Middlehurst, Eds., Chicago, University of Chicago Press, 272-342.

Hayes, D. S., 1967: An absolute calibration of the energy distribution of twelve spectrophotometric standard stars. Ph.D. thesis, University of California at Los Angeles.

Herzberg, G., 1950: Infrared and Raman Spectra. Princeton, D. Van Nostrand Co., Inc., Chap. 4.

Hunten, D. M., 1972: The atmosphere of Titan. Comments on Astrophys. Space Phys., 4, 149-154.

Hunten, D. M., 1973a: The escape of H_2 from Titan. J. Atmos. Sci., 30, 726-732.

Hunten, D. M., 1973b: The escape of light gases from planetary atmospheres. J. Atmos. Sci., 30, in press, Nov. 1973.

Hunten, D. M., and Strobel, D. F., 1974: Production and escape of terrestrial hydrogen. J. Atmos. Sci., 31, in press, Jan. 1974.

Jeffreys, H., 1954: Second-order terms in the figure of Saturn. R. Astron. Soc., Mon. No., 114, 433.

Johnson, H. L., 1965: The absolute calibration of the Arizona photometry. Comm. Lunar and Planetary Lab., 3, No. 53, 73.

Johnson, H. L., 1970: The infrared spectra of Jupiter and Saturn at 1.2-4.2 microns. Ap. J. Letters, 159, 1.

Joyce, R. R., Knacke, R. F., and Owen, T., 1973: An upper limit on the 4.9-micron flux from Titan. Ap. J. (Letters), 183, L31.

Khare, B. N., and Sagan, C., 1973: Red clouds and reducing atmospheres. Icarus, 20, in press.

Kovalevsky, J., 1970: Determination des masses des planètes et satellites, in Surfaces and Interiors of Planets and Satellites. A. Dollfus, Ed., London, Academic Press.

Kuiper, G. P., 1944: Titan: A satellite with an atmosphere. Ap. J., 100, 378-383.

Kuiper, G. P., 1952: Planetary atmospheres and their origin, in The Atmospheres of the Earth and Planets. G. P. Kuiper, Ed., Chicago, University of Chicago Press, 306-405.

Labs, D., and Neckel, H., 1967: The absolute radiation intensity of the center of the solar disk in the spectral range $\lambda3288$ to $\lambda12480$Å. Z. Astrophys., 65, 131.

Leovy, C. G., and Pollack, J. B., 1973: A first look at atmospheric dynamics and temperature variations on Titan. Icarus, 19, 195.

Lewis, J. S., 1971: Satellites of the outer planets: Their physical and chemical nature. Icarus, 15, 174-185.

Lewis, J. S.: 1972, Low temperature condensation from the solar nebula, Icarus, 16, 241.

Lewis, J. S., 1973: Composition of Planets and Satellites. Submitted to Sci. Amer..

Lewis, J. S., 1973: Chemistry of the outer solar system, Space Sci. Rev., 14, 401.

Lewis, J. S., and Prinn, R. G., 1973: Titan revisited. Comments on Astrophys. Space Phys., 5, 1-7.

Light, E. S., and Danielson, R. E., 1973: Further analysis of the limb darkening curves of Uranus. Bull. A. A. S., 5, 291.

Low, F. J., 1965: Planetary radiation at infrared and millimeter wavelengths. Lowell Observatory Bull., 6, no. 128, 184-187.

Low, F. J., and Armstrong, K. R., 1973: Effective temperatures and infrared continua of the planets and satellites. Bull. A. A. S., 5, 306.

Lyot, B., 1929: Research on the polarization of light from planets and from some terrestrial substances. NASA Technical Translation, F-187.

Lyot, B., 1953: Planetary observations at the Pic du Midi. Bull. Astron., 67, 31.

Macy, W. W., 1973: Ph.D. thesis, Princeton University.

McCord, T. B., Johnson, T. V., and Elias, J. H., 1971: Saturn and its satellites: Narrow-band spectrophotometry (0.3-1.1 µm). Ap. J., 165, 413-424.

McDonough, T. R., and Brice, N. M., 1973a: New kind of ring around Saturn? Nature, 242, 513.

BIBLIOGRAPHY AND REFERENCES   (CONTD)

McDonough, T. R., and Brice, N. M., 1973b: A Saturnian gas ring and the recycling of Titan's atmosphere.   Icarus, 20, 136.

McGovern, W. E., 1971:   Upper limit of hydrogen and helium concentrations of Titan, in Planetary Atmospheres.   C. Sagan, T. C. Owen, H. J. Smith, Eds., I. A. U. Symposium 40, Dordrecht-Holland, D. Reidel Publishing Co., 394-400.

Message, P. J., 1972:   A survey of dynamical data for the major planets and satellites.   Phys. Earth Planet. Interiors, 6, 17-20.

Moroz, V. I., 1966:   The spectra of Jupiter and Saturn in the 1.0-2.5 μm region. Soviet Astron., 10, 457.

Morrison, D., 1973:   New techniques for determining sizes of satellites and Asteroids, Comments on Astrophys. and Space Phys., 5, 51.

Morrison, D., Cruikshank, D. P., and Murphy, R. E., 1972:   Temperatures of Titan and the Galilean satellites at 20 microns.   Ap. J. Letters, 173, 143-146.

Munch, G., 1973:   On the spectrum of Titan.   Bull. A. A. S., 5, 305.

Noland, M., Veverka, J., Morrison, D., Cruikshank, D. P., Lazarewicz, A., Elliot, J., Goguen, J., and Burns, J., 1973:   Six color photometry of Iapetus, Titan, Rhea, Dione, and Tethys.   In preparation.

O'Leary, B., 1972:   Frequencies of occultation of stars by planets, satellites, and asteroids.   Science, 175, 1108.

O'Leary, B., and van Flandern, T. C., 1972:   Io's triaxial figure.   Icarus, 17, 209.

Pickering, E. C., 1913:   Harvard Bulletin, No. 538.

Pollack, J. B., 1973:   Greenhouse models of the atmosphere of Titan.   Icarus, 19, 43.

Sagan, C., 1971:   The solar system beyond Mars: An exobiological survey. Space Sci. Reviews, 11, 73.

Sagan, C., 1973:   The greenhouse of Titan.   Icarus, 18, 649-656.

Strobel, D. F., 1973:   The photochemistry of hydrocarbons in the Jovian atmosphere.   J. Atmos. Sci., 30, 489-498.

Strobel, D. F., and Smith, G. R., 1973:   On the temperature of the Jovian thermosphere.   J. Atmos. Sci., 30, 718-725.

Sullivan, R. J., 1973:   Titan: A model for a toroidal gas cloud surrounding its orbit.   Preprint.

Tabarié, N., 1973: Atomic hydrogen distribution in Titan's atmosphere. Service d'Aeronomie du CNRS, Report No. 61 G 1973.

Taylor, G. E., 1972: The determination of the diameter of Io from its occultation of β Scorpii C on May 14, 1971, Icarus, 17, 202.

Taylor, G. E., 1973: Private communications.

Thorndike, A. M., 1947: The experimental determination of the intensities of infrared absorption bands. III. Carbon dioxide, methane, and others. J. Chem. Phys., 15, 868-874.

Trafton, L. M., 1972: Newly discovered absorptions in Titan's infrared spectrum. Bull. A. A. S., 4, 367.

Trafton, L. M., 1972a: On the possible detection of H_2 in Titan's atmosphere. Ap. J., 175, 285-293.

Trafton, L. M., 1972b: The bulk composition of Titan's atmosphere. Ap. J., 175, 295-306.

Trafton, L. M., 1973: Interpretation of Titan's infrared spectrum in terms of a high-altitude haze layer. Bull. A. A. S., 5, 305.

Trafton, L. M., 1973a: Titan: Unidentified strong absorptions in the photometric infrared. Submitted to Icarus.

Turekian, K. K., and Clark, S. P. Jr., 1969: Inhomogeneous accumulation of the Earth from the primitive solar nebula. Earth Planet. Sci. Letters, 6, 346.

Veverka, J., 1970: Photometric and polarimetric studies of minor planets and satellites. Ph.D. thesis, Harvard University.

Veverka, J., 1973: Titan: Polarimetric evidence for an optically thick atmosphere? Icarus, 18, 657-660.

Veverka, J., Wasserman, L., Elliot, J., Sagan, C., and Liller, L., 1973: The occultation of Beta Scorpii by Jupiter. I. The structure of the Jovian upper atmosphere. Astron. J., in press.

Wasserman, L., and Veverka, J., 1973a: Analysis of spikes in occultation curves: A critique of Brinkmann's method. Icarus, 18, 599.

Wasserman, L., and Veverka, J., 1973b: On the reduction of occultation light curves. Icarus, in press.

Whitehill, L., 1971: Private communication.

Willstrop, R. V., 1960: Absolute measures of stellar radiation. Monthly Notices Roy. Astron. Soc., 121, 17.

Woltjer, J., 1928: The motion of Hyperion. Ann. d. Sterrewacht Leiden, 16, part 3.

Younkin, R. L., 1970: Spectrophotometry of the Moon, Mars and Uranus. Ph.D. thesis, University of California at Los Angeles.

Zellner, B., 1973: The polarization of Titan. Icarus, 18, 661-664.